PHYSICS RESEARCH AND TECHNOLOGY

RECENT RESEARCH IN QUANTUM GRAVITY

PHYSICS RESEARCH AND TECHNOLOGY

Additional books in this series can be found on Nova's website under the Series tab.

Additional E-books in this series can be found on Nova's website under the E-book tab.

PHYSICS RESEARCH AND TECHNOLOGY

RECENT RESEARCH IN QUANTUM GRAVITY

ARUNDHATI DASGUPTA
EDITOR

nova publishers
New York

For permission to use material from this book please contact us:
Telephone 631-231-7269; Fax 631-231-8175
Web Site: http://www.novapublishers.com

Additional color graphics may be available in the e-book version of this book.

Library of Congress Cataloging-in-Publication Data

Recent research in quantum gravity / [edited by] Arundhati Dasgupta.
p. cm.
Includes bibliographical references and index.
ISBN 978-1-61942-386-2 (hardcover)
1. Quantum gravity. I. Dasgupta, Arundhati.
QC178.R43 2011 530.14'3--dc23 2011050198

Published by Nova Science Publishers, Inc. † New York

CONTENTS

Preface		**vii**
Chapter 1	Introduction *Arundhati Dasgupta*	**1**
Chapter 2	Directions in Causal Set Quantum Gravity *Sumati Surya*	**7**
Chapter 3	The Asymptotic Safety Program in Quantum Einstein Gravity *Martin Reuter and Frank Saueressig*	**41**
Chapter 4	Midi-Superspace Models for Quantum Black Holes *J. Gegenberg and G. Kunstatter*	**95**
Chapter 5	Is the Spacetime Metric Euclidean Rather than Lorentzian? *Rafael D. Sorkin*	**115**
Chapter 6	Euclidean Quantum Gravity: Recent Results *Arundhati Dasgupta*	**137**
Index		**169**

PREFACE

The book has a collection of articles on quantum gravity. A common thread connecting the articles is the quest to discover some new aspect of the ultimate theory of nature. Two new approaches to quantum gravity: The Causal Set Theory and The Asymptotic Safety Program have been described in a pedagogical way. Gravity is the theory of space-time, and the quantum theory of gravity seeks to describe space-time at the microscopic scale. Experiments have not yet probed or resolved distances at which the physics of space-time might be `quantised' or discrete as the same way matter was found to be made up of atoms. The causal set theory is a very radical approach to the subject using posets or partially ordered sets to describe a coarse-grained space-time. The asymptotic safety scenario is another promising new approach which has given us new insights into short distance physics showing it to be fractal in nature. The book also includes three other articles about path-integrals and 2-dimensional gravity. All the articles are written in a accessible way and thus this book should prove useful to physicists begining their research in quantum gravity. It is also a useful reference for advanced researchers.

I would like to thank the invited contributers of the articles Sumati Surya, Martin Rueter and Frank Saueressig, Gabor Kunstatter and Jack Gegenberg, and Rafael Sorkin for their efforts and the excellent articles.

The publication of this book got delayed by almost three years due to the birth of my daughter, and this book is dedicated to all the working mothers of the world many of whom do not yet have the luxury of parental leave or family support. I would like to thank my family for their support.

In: Recent Research in Quantum Gravity
Editor: A. Dasgupta, pp. 1-5
ISBN: 978-1-61942-386-2

Chapter 1

INTRODUCTION

Arundhati Dasgupta
Department of Physics and Astronomy, University of Lethbridge,
Lethbridge, Canada

Quantum Gravity is a puzzle we have not solved yet. Several pieces of the puzzle have been pieced together, but remaining ones are proving tough to arrange. The nature of electric, weak and strong forces of particle physics had been revealed to us through experimental observations. We simply rebuilt the natural picture using mathematical and theoretical physics tools known as the standard model of particle physics. In contrast, nature has not given us even one clue about quantum gravity. There are no observational signatures of quantum gravity, yet we have the mathematical or theoretical tools which we hope will cohere into a meaningful picture. We also hope that this theory will unite with the standard model of particle physics to provide one theory for the entire dynamics of nature.

For decades physicists have been declaring that we have solved and understood all natural phenomena, but that arrogance is simply misplaced. Currently we have observational evidence of 4% of the Universe and 96% of the Universe is dark matter and dark energy or simply unknown. However, given the existing laws of physics quantum gravity is a natural conclusion to the model of the Universe we have painstakingly built through thousand of years of human history. The rational explanation of motion in nature started with Newton and indeed the entire classical dynamics is due to just four types of interactions, the electric, the strong and weak nuclear forces and gravity. Of the four, the electric and the gravitational forces are observed and perceived in day to day activities. The strength of the gravitational interaction is very weak though for massive bodies and astrophysical scales this interaction dominates. At atomic scales $\sim 10^{-9}$ cm the electric interaction's quantum manifestation leads to the formation of the atom. As the gravitational force is much weaker, its quantum nature should emerge at 10^{-33} cm or the Planck length (l_p). Our accelerators have not resolved such distances yet though theory has tried to predict physics at such length scales. Unfortunately this human endeavour has not proved easy, and the emergent physics is difficult to interpret. Theoretically we have very clear methods of quantising a classical theory. Quantising gravity should be a straightforward task given the experimentally verified classical theory of the dynamics of space-time, the General theory of relativity. At l_p scales, space-time might have a 'discretised structure' just as atoms are the building blocks of matter. However, the straight forward methods lead to infinities or are mired in technical complications. Most of 'quantum interaction' in the standard model is obtained by quantising a free Lagrangian and then introducing interactions as fluctuations or perturbations to the free propagation. These perturbative methods fail for gravity and it was plausible that if we extended the methods of quantisation to include non-perturbative methods the quantum gravity amplitude might be computable. Indeed recently all the existing theoretical pieces of the puzzle to the ultimate theory of nature have been obtained using non-perturbative methods. Lattice Gravity and non-perturbative quantisation using loop quantum gravity have given results which reveal the nature of space-time and its behaviour at Planck length scales. In one description it emerges as fractal with the Hausdorff dimension as 2 at short distances, in the other area is a integer multiple of the fundamental bit $8\pi l_p^2$. However in the absence of experiments, none of the above can be verified. Yet the search must be continued through development of new experimental methods and indirect signatures of quantum gravity in current day observable phenomena.

One of the compelling reasons for the search of a quantum theory of gravity is the presence of pathological singularities in classical solutions. The cosmological solutions which describe the evolution of the universe from the time of its formation has a curvature singularity from which suddenly space-time appears in the form of a 'big bang'. The question remains is what exactly is the

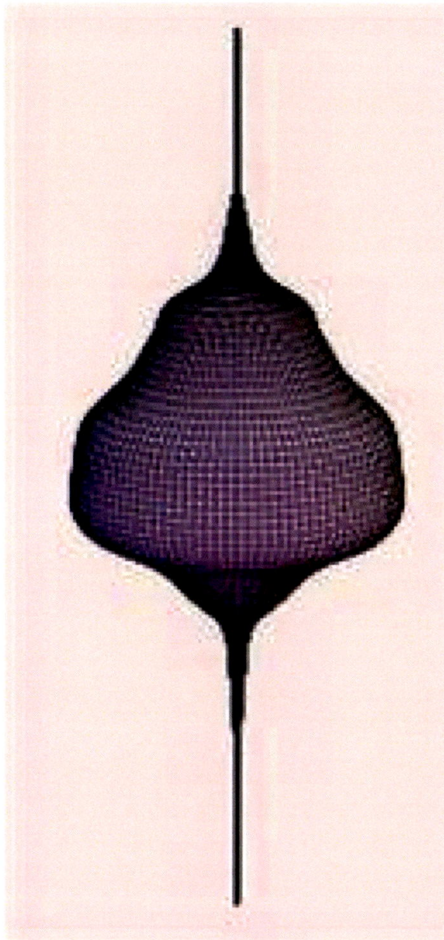

Figure 1. The emergent space-time obtained from CDT copyright J. Ambjorn, J. Jurkiewicz, R. Loll.

physics of this singularity and what precedes this? The black holes are strong gravitating objects which absorb everything including light, and at their center sits a singularity. All matter which falls into a black hole is eventually destroyed at the central singularity, and this violates the principle of conservation of mass and energy which is so sacrosant in the laws of physics. There has been no experimental observations of these singularities, and what we have observed seems to exclude such pathological and destructive behaviour of nature. Thus a correct description of nature should avoid these pathologies. Quantum theory has previously come to the rescue by smoothening out singularities, e.g. obtaining an atomic structure which stabilises the infinity of electron-proton potential as predicted by Maxwell's classical laws. It is hoped that if we find the correct theory of quantum gravity, these singularities will not be there.

Using pure theoretical methods there are various approaches, namely Loop Quantum Gravity (LQG), Causal Dynamical Triangulation (CDT), String Theory (ST) which have acquired importance in the physics community. In this book, two other approaches: The Causal Set Theory and The Asymptotic Safety Program are descibed in great pedagogical detail. These approaches are yet to obtain the popularity of the LQG, CDT and ST approaches to quantum gravity.

The Causal Set approach is described in *Directions in Causal Set Quantum Gravity'* by Sumati Surya (Chapter 2). A Lorentzian space-time is described by partially ordered sets (posets) and the structure of space-time emerges from the stochastic description of the posets. This approach to quantum gravity is rooted in the 'entropic' description of space-time forced by the presence of horizons. This approach is promising particularly at this time as there are attempts to reformulate '*classical gravity*' as dynamics of entropic bits [4].

The asymptotic safety program is described in the article 'The Asymptotic Safety program in Quantum Einstein Gravity' by Martin Reuter and Frank Saueressig (Chapter 3). This approach postulates a renormalisation flow for gravity and obtaines a very interesting result: the space-time flows to a 2-dimensional space-time at short length scales! This remarkable result was lated confirmed by the CDT calculations and thus there might be a truth to the fractal nature of short distance physics. Once again it is only experiment which can reveal the truth. There have been further developments in this approach and the reader can find recent work described in [5].

The article on 'Midi-Super space models of quantum black holes' by Jack Gegenberg and Gabor Kunstatter (Chapter 4) gives an example of a case where the quantisation problem of a blackhole has been mapped to a quantum mechanical system. It uses a non-standard quatisation of the reduced Hamiltonian and shows us new methods. The use of alternative quantisations to solve the existence of black hole thermodynamics also makes this article an asset to this collection.

The next two articles 'Is the space-time metric Euclidean rather than Lorentzian?' by Rafael Sorkin (Chapter 5) and 'The measure in Euclidean quantum gravity' by myself (Chapter 6) provide interesting perspective on the continuum path-integral quantisation. It is interesting to see that new physics and observations still exist in a approach physicists had assumed has reached a closed end and the search for alternative quantisations had started. In the article 'The measure in Euclidean quantum gravity' a brief description of the CDT quantisation is given. This appraoch has recently shown very promising results, including an emergent universe. Figure (1) is such a simulation, with 3-volume plotted in time (the vertical axis). It resembles a semiclassical universe with a positive cosmological constant.

In summary the collection of articles are very original and provide new insights in the field of quantum gravity. It is very difficult to say if any of the described theories will emerge as the quantum theory of gravity. However each of this gives some clue to what the theory of quantum gravity might be.

References

[1] T. Thiemann, *Modern Canonical Quantum General Relativity* Cambridge University Press (2007).

[2] J. Ambjorn, A. Gorlich, J. Jurkiewicz, R. Loll *CDT -an Entropic Theory of Quantum Gravity*, Lectures given at International Workshop on Lattice and Continuum approaches to Quantum Gravity, Breighton UK. arXiv:1007.2560 [hep-th].

[3] O. Aharony. S. Gubser, J. Maldacena, H. Ooguri, Y. Oz Large N field theories, string theory and gravity, *Phys. Rept.* 323 183-386 (2000).

[4] E. Verlinde, *On the Origin of Gravity and the Laws of Newton'* arXiv: 1001.0785 [hep-th].

[5] U. Harst, M. Reuter, *QED coupled to QEG* arXiv:1101.6007 [hep-th].

In: Recent Research in Quantum Gravity
Editor: A. Dasgupta, pp. 7-40
ISBN: 978-1-61942-386-2

Chapter 2

DIRECTIONS IN CAUSAL SET QUANTUM GRAVITY

Sumati Surya
Raman Research Institute, Bangalore, India
Department of Physics, McGill University, Montreal, Canada

Abstract

The last decade has seen a steady growth of work in causal set quantum gravity. In this approach, the spacetime continuum arises as an approximation to a fundamentally discrete substructure, the causal set, which is a locally finite partially ordered set. The causal set paradigm was elucidated in a classic paper by Bombelli, Lee, Meyer and Sorkin in 1987 [1]. While early kinematical results already showed promise, the program received a substantial impetus with the work of Rideout and Sorkin on a classical stochastic growth dynamics for causal sets [2]. Considerable progress has been made ever since in uncovering the hidden structures of causal set theory, leaving undisturbed the basic paradigm set out in [1]. Recent highlights include a causal set expression for the Einstein-Hilbert action [3, 4] and the construction of a scalar field Feynman propagator on a fixed causal set [5, 6]. The aim of the present article is to give a broad overview of the results in causal set theory while pointing out directions for future investigations.

1. Introduction

The origins of the causal set paradigm lie in the rich soil of Lorentzian geometry. Unlike a Riemannian space, a Lorentzian spacetime (M, g) possesses an additional structure, the *causal structure* $(M, \prec)$ where for a pair of events $x, y \in M$, $x \prec y$ if x is to the causal past of y [7, 8]. For a causal spacetime, i.e., one which has no closed causal curves, $(M, \prec)$ is a partially ordered set (or poset), namely $\prec$ is

1. Reflexive: $x \prec x$.

2. Acylic: $x \prec y$ and $y \prec x$ implies that $y = x$.

3. Transitive: $x \prec y$ and $y \prec z$ implies $x \prec z$.

In addition to the causal relation $\prec$ one also has the chronological relation $\prec\!\!\prec$ which satisfies both acyclicity and transitivity but not reflexivity[1]. Viewed in terms of the partial order $(M, \prec)$, M is merely the set of spacetime events, stripped of its standard manifold-like topological character.

Spacetime causal structure was studied extensively in the 60's and 70's and led to a deeper understanding of the fundamentally Lorentzian features of spacetime like black holes and their associated singularities [7, 8, 10, 11, 12, 13]. These investigations suggested that causal structure plays a primitive, rather than a derivative role in Lorentzian geometry leading naturally to the question – can spacetime topology and geometry be derived from the poset $(M, \prec)$?

The answer to the former question has been known for several decades now, when restricted to the class of strongly causal spacetimes [7, 8]. First, notice that the relational connectivity of a generic poset ensures that it contains non-trivial topological information – indeed, posets admit a host of different non-trivial topologies [14, 15]. Of these, the Alexandroff topology $\mathcal{A}$ generated from a basis of Alexandroff intervals $I(p, q) \equiv I^+(p) \cap I^-(q)$, where $I^+(p) \equiv \{s | p \prec\!\!\prec s\}$ and $I^-(q) \equiv \{s | s \prec\!\!\prec q\}$, is of particular interest. It is not difficult to prove that in strongly causal spacetimes $\mathcal{A}$ is homeomorphic to the manifold topology $\mathcal{M}$, or equivalently, $\mathcal{A}$ is homeomorphic to $\mathcal{M}$ iff $\mathcal{A}$ is Hausdorff [10]. It was recently demonstrated in [9] that manifold topology can also be

[1]This relation can be derived from $\prec$ when the spacetime is either past or future distinguishing [9].

obtained from a topology constructed from the causal structure in a more general class of spacetimes. Thus, manifold topology can indeed be obtained from a causal topology on $(M, \prec)$, for very general classes of spacetimes. This is somewhat startling at first glance since the standard manifold topology for M is intrinsically Riemannian, being constructed from a basis of open balls of varying radii, the latter being defined with respect to a Riemmanian distance. On the other hand, any topology on $(M, \prec)$ is constructed, by design, from intrinsically Lorentzian structures.

That there is a relation between geometry and causal structure is perhaps not altogether surprising: it is standard knowledge that the causal structure is unchanged under conformal transformations of the spacetime geometry. Using results due to Hawking, King and McCarthy [13] Malament [16, 17] showed that causal bijections (i.e., maps that preserve the causal structure) between 4-dimensional future and past distinguishing spacetimes are also conformal isometries. In other words, if a pair of 4-dimensional spacetimes admit a bijective map that preserves the causal structure then these spacetimes are in fact conformally related. It is possible to show that causal bijections cannot exist between spacetimes with different dimensions ($d > 2$) and hence causal structure also determines the spacetime dimension [9]. These results imply that the conformal geometry is completely determined by the causal space $(M, \prec)$ provided the latter satisfies certain causality conditions. In a 4-dimensional spacetime the causal structure is therefore "$9/10$th" of the spacetime geometry and the remaining "$1/10$" is determined by the local volume element so that [18, 19]

$$\mathrm{GEOMETRY} = \mathrm{CAUSAL\ \ STRUCTURE} + \mathrm{VOLUME} \qquad (1)$$

This is one of the starting points for the causal set paradigm.

The second ingredient that goes into causal set theory comes from the quantum world, and its conjunction with gravity. This suggests that there is a fundamental Plank scale cut-off l_p in spacetime – a suggestion that is reinforced by the apparent "quantisation" of the black hole area law[2]. How this fundamental cut-off is implemented as a "discretisation" of spacetime is of course dependent on what other continuum baggage one wishes to retain. Rather than employing a Lorentz violating cut-off like the Plank energy (or *length* or time), it is desirable

[2]An adage attributed to Mark Kac "Be Wise – Discretise" while applicable to various physical and mathematical situations, seems particularly appropriate to quantum gravity!

to use the covariantly well-defined Plank spacetime *volume* $V_p = l_p^4$ as a cut-off. Introducing such a cut-off in the poset $(M, \prec)$ produces a *causal set* C, which apart from being a poset is also "locally finite". Namely, if $J^+(x) \equiv \{y|x \prec y\}$ and $J^-(x) \equiv \{y|y \prec x\}$, then the interval $J(x,y) = J^+(x) \cap J^-(y)$ is required to be of finite cardinality. The prescription of local finiteness formalises the idea of a fundamental spacetime discreteness – spacetime regions that have finite volume should contain only a finite number of fundamental spacetime atoms.

Indeed, it is precisely this added ingredient of discreteness which makes it possible to reconstruct, in an appropriate sense, the full spacetime geometry from the causal set in the spirit of Statement (1). Normalising the volume with respect to a volume cut-off, this suggests the natural correspondence

$$\begin{aligned} \text{Order} &\rightarrow \text{Causality} \\ \text{Number} &\rightarrow \text{Volume}. \end{aligned} \tag{2}$$

The particulars of course depend on the desired nature of the discrete-continuum correspondence. Importantly, if the discrete substratum is to be thought of as fundamental, then the continuum can arise only as an approximation. The analogy to keep in mind is that of a fluid whose apparently continuum nature has an underlying, discrete, molecular structure. In lattice gauge theory, or Causal Dynamical Triangulations [20], in contrast, the continuum is regarded as fundamental and is approximated by the discrete lattice. In this case, the continuum limit is essential, since it represents the physical limit of a discrete approximating scheme. However, it is precisely because the continuum is not a fundamental construction in the causal set approach that the continuum *limit* plays no fundamental role in the theory. As in statistical physics, one is therefore interested in seeing how coarse grained continuum structures emerge from the underlying causal set.

However, to decide on the rules for when the approximation is a good one, it is useful to start with the reverse question: what is the causal set that underlies a continuum spacetime? The naivest such discretisation of Minkowski spacetime is via a regular lattice of the sort depicted in Fig 1. In the figure on the left the spacetime interval depicted has a volume $V = 4V_c$ and contains precisely 4 elements of the lattice. However, as shown in the figure on the right, under a boost this same region contains no lattice elements[3]. Thus, the second correspondence in Statement (2) is not consistent with a regular lattice, no matter how clever the construction. The only discretisation that can implement

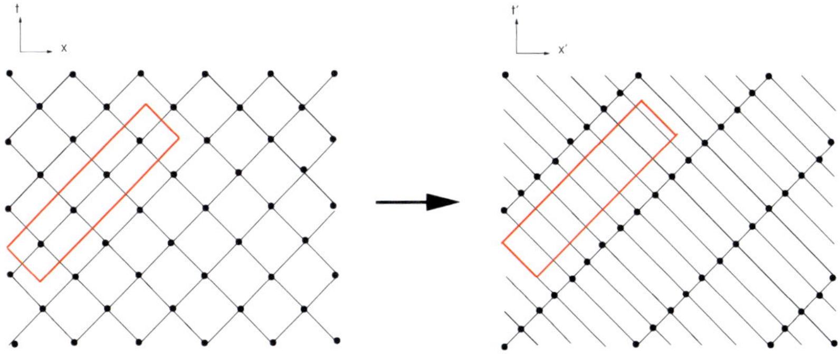

Figure 1. Diamond lattice under a boost. A fixed volume region containing 4 elements before the boost becomes empty after the boost.

the spirit of this correspondence is a "random lattice" [21]. A causal set C can be generated from a continuum spacetime (M, g) via a random sprinkling of points ordered by the induced causal order. The most appropriate (and possibly unique) random distribution for our purposes is the Poisson distribution $P_V(n) = (1/n!)\exp^{-V/V_c}(V/V_c)^n$, where $P_V(n)$ is the probability of finding n elements in a spacetime region of volume V for a given cut-off V_c. For this distribution, $\langle n \rangle = V/V_c$ which means that the number to volume correspondence holds on average, thus satisfying the spirit of the Malament theorem. Thus, one has the causal set maxim as stated by Rafael Sorkin [18]

$$\text{ORDER} + \text{NUMBER} \approx \text{SPACETIME GEOMETRY} \tag{3}$$

as a discrete analogue of (1). Fig 2 depicts a causal set obtained by a random discretisation of a region of 2d Minkowski spacetime. Indeed, a random lattice is the best that one can do to regain the number to volume correspondence and while this leads to several technical complications when extracting continuum information from a causal set, it has the added bonus of preserving local Lorentz

[3]An equivalent, passive viewpoint, is the observation that most intervals of spacetime volume $V = 4V_c$ in the figure on the left contain no lattice elements.

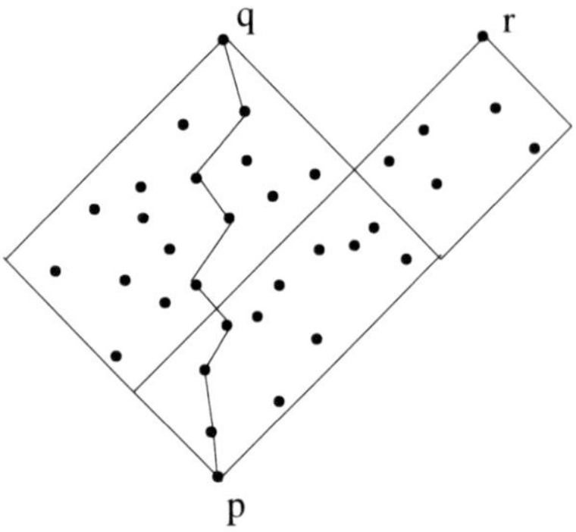

Figure 2. Two Alexandroff intervals $J(p,q)$ and $J(p,r)$ in a causal set that is approximated by a region of 2d Minkowski spacetime. The length of the longest chain from p to q gives rise to the time-like distance between p and q.

invariance [22]. This in turn has a profound significance on the phenomenology associated with causal sets.

We have described how to obtain a causal set which underlies a continuum spacetime but the more fundamental question we seek to answer is how the continuum emerges as an approximation of the theory. Given a causal set C, we will call an embedding $\Phi : C \to (M,g)$ *order preserving* if the order relation in C is mapped bijectively to the causal order induced on $\Phi(C)$. It is an important fact that not all causal sets can be embedded via an order preserving map into a given spacetime [23]. In addition, the embedding Φ is *faithful* if it is order preserving and $\Phi(C) \subset M$ is a high probability Poisson distribution in (M,g) [24]. A causal set C is then said to be approximated by a continuum spacetime (M,g) if C can be faithfully embedded into (M,g). Note that the continuum spacetime that one obtains via the approximation cannot be too detailed in its description — structure (both topological and geometrical) on scales at or smaller than the Planck scale cannot be encoded in the causal set. It is thus important for a faithful embedding to be approximately unique – a given causal set should not approximate to spacetimes that differ on scales much larger than the discreteness scale. If this were the case, then it means that spacetimes cannot be distinguished solely by the underlying casual set. A Hauptvermutung or fundamental conjecture of causal set theory states

Hauptvermutung: *If a causal set C faithfully embeds at the same density into two distinct spacetimes (M_1, g_1) and (M_2, g_2) then these spacetimes are related by an approximate isometry.*

Here, an approximate isometry captures the notion that the two spacetimes differ only at scales close to V_c. The conjecture has been proved for the case when the locally finite condition is relaxed to locally countable and dense [25]. The conjecture has however not been proved in general and this remains an important open problem in causal set theory. Much of the difficulty lies in defining the closeness of two Lorentzian geometries, although progress has been made in this direction [26, 27]. On the other hand, a considerable body of circumstantial evidence has accumulated over the years in support of the conjecture from the construction of continuum geometrical and topological information from the causal set and we will describe this "kinematics" in Section 2 In Section 3 we discuss phenomenological predictions that arise from causal sets. The most important of these is the prediction of the value of the cosmological constant [28]. In Section 4 we discuss both the classical and quantum dynamics of causal sets and end with some concluding remarks in Section 5

The aim of this article is not to attempt an exhaustive overview of causal sets nor is it to paint a picture of a completed theory in which most fundamental questions have been addressed. Instead, it is to give a basic outline of the program and point out the directions that it is currently taking. In the process it will become possible to describe in some detail certain important open questions, and the difficulties one might encounter in trying to answer them. The outlines or axioms of the approach are clear and hence offer useful constraints on how to proceed. While considerable progress has been made in the last decade and new directions have been opened up, there are still several important questions that need attention.

2. Kinematics

In this section, we will focus our attention on the question, "Where is the geometry and topology of a continuum spacetime (M, g) encoded in a causal set C?" Finding answers to this query requires a sometimes laborious and sometimes inspired search for continuum-like structures within the causal set. Every successful identification of such a continuum property is a confirmation of the

Hauptvermutung since the same causal set C cannot then embed into a spacetime (M', g') with a significantly different continuum property.

Importantly, standard poset theory does not always provide ready answers to this question– while a poset itself admits several natural geometric and topological structures, these are not necessarily appropriate in the continuum approximation. The main difficulty arises from the randomness of the discretisation. A useful contrast is a Regge-discretisation [29] which uses a simplicial decomposition – not only are topological structures retained under the discretisation, but there is a simple procedure for calculating the local curvature using nearest neighbours. In a causal set the nearest neighbour of an element p is one that is linked to it: $p \prec q$ is said to be a link if for every r with $p \prec r \prec q$, either $r = p$ or $r = q$. A link is thus an irreducible relation in a causal set since it cannot be deduced from transitivity. For a causal set that is approximated by Minkowski spacetime, every element therefore has an infinite number of nearest neighbours extending all along its future and past light-cones (see Fig 3). This points to an inherent *non-locality* in causal sets, a legacy of discreteness combined with local Lorentz invariance [30, 31]. This non-locality makes it non-trivial to construct familiar continuum quantities and it is a happy fact that, despite this, much progress has been made.

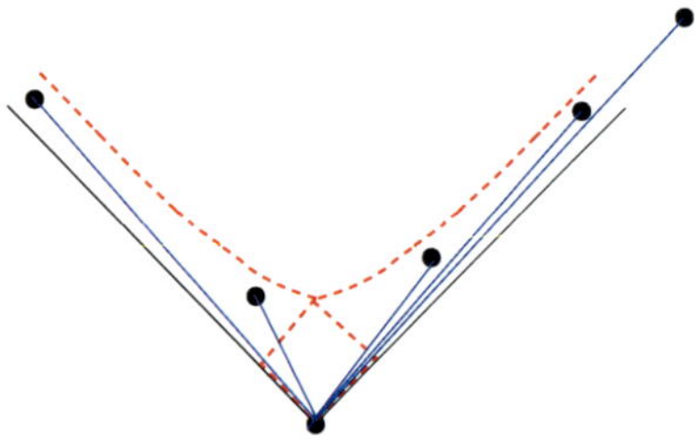

Figure 3. The links from an element in a causal set that is approximated by flat spacetime lies all along the null cones. The figure shows a few (of the infinite) future directed links.

The most rudimentary topological property of a spacetime is its dimension. Let C be a causal set which faithfully embeds into a flat spacetime interval $J(p,q)$ of dimension d and volume V. The Myrheim-Meyer dimension estimator for this causal set uses the distribution of "chains". A chain c is a totally ordered subset of C, i.e., for every pair of elements $x, y \in c$, either $x \prec y$ or

$y \prec x$ (Fig 4). It was shown in [23] that for $V_c \ll V$, the average number of k-element chains $\langle C_k \rangle = f(d,k)\langle N \rangle^k$, where $\langle N \rangle = (V/V_c)$ and $f(d,k)$ is an explicit function of d and k. This function can then be inverted to obtain the dimension. For $k = 2$, this estimator is called the Myrheim-Meyer dimension [32], where $\langle C_2 \rangle$ is simply the average number of relations in C. $f(d,2)$ is then half of the so-called "ordering fraction" of C, i.e., the ratio of the number of relations in C to the number of possible relations between n elements.

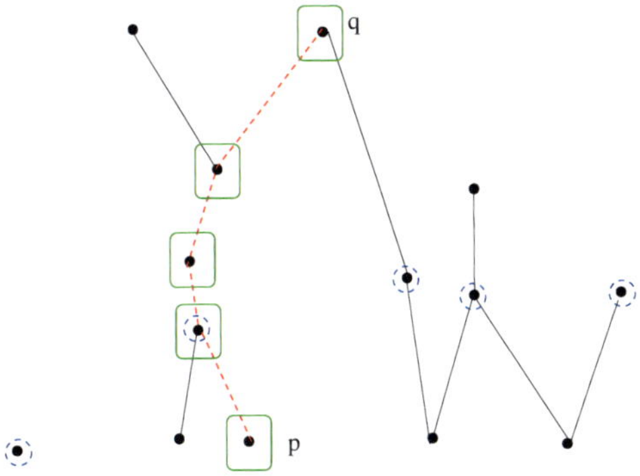

Figure 4. A link, or Hasse diagram of a 14 element causal set, where only the links are shown. The relations can deduced from transitivity. We show an example of a 3-element chain marked by large red circles and a 5-element antichain marked by small dashed blue circles. The elements in the Alexandroff interval $J(p,q)$ are marked by green boxes.

Another dimension estimator is the mid-point scaling dimension, which uses the relationship between the volume V of a spacetime interval $J(p,q)$ in $\mathbb{M}^d$ and the length τ of the time-like geodesic from p to q, i.e., $2^d = V/V'$, where the midpoint $m \in J(p,q)$ is the one that maximises the volume V' of the smaller of the two intervals $J(p,m)$ and $J(m,q)$ [19, 33]. Numerical tests show that these dimension estimators reproduce the continuum dimension fairly accurately for causal sets that approximate to flat spacetimes for small d [23, 34].

In [23] an analysis of $\langle C_k \rangle$ for causal sets that approximate to both anti-deSitter and deSitter spacetimes suggested that not only dimension, but also the Gaussian curvature and the height of the conformally related flat interval could be obtained by solving a system of three equations. The numerical results though promising, were not entirely conclusive since relatively small causal sets were used. In [34] a procedure was proposed for finding Minkowski-like intervals in a class of conformally flat spacetimes from which to calculate the spacetime dimension, and numerical simulations were carried out to verify the procedure. For large enough causal sets these simulations work very well for the Myrheim-Meyer dimensions [34].

While these two dimensional estimators do seem to yield the correct values for Minkowski spacetime, it is an important open question whether and how they can be extended to arbitrary curved spacetimes. It is tempting to think that these results should still be valid since dimension is locally defined in the continuum. Indeed, in the continuum every event lies in a neighbourhood which is diffeomorphic to flat spacetime. But how does one identify this special neighbourhood in a causal set when there is no intrinsically Lorentzian definition of such a neighbourhood?

The continuum result that the Alexandroff topology is the manifold topology in strongly causal spacetimes suggest that this is the most natural topology for a causal set. While this will give rise to the manifold topology in the continuum limit, what is of relevance to causal sets is not the fine manifold topology but some coarser sub topology. In [35] it was suggested on general grounds that it may suffice to consider a finitary topology generated by locally finite covering of a space. In Riemannian spaces such locally finite sub topologies are easy to construct from a discretisation. For example, the circumspheres of a Delaunay triangulation of a space provide a locally finite sub topology. However, Alexandroff intervals do not provide locally finite coverings – every element of the causal set is contained in an infinite number of such intervals in Minkowski spacetime.

In order to construct a useful locally finite covering, one can implement a localisation via the causal set analog of a spacelike hypersurface. This is an inextendible antichain A or a maximal collection of unrelated elements in C (see Fig 4). Since a choice of A is a choice of frame, one must also sample over a sufficiently large ensemble of randomly chosen inextendible antichains. This was the strategy adopted in [36, 37] for constructing the homology of a causal set which is approximated by a globally hyperbolic spacetime. Since A itself carries only the discrete topology, it needs to borrow information from the larger causal set. This is done via a "volume" thickening $T_n(A) \equiv \{x|x \succ A, |J(A,x)| \leq n\}$, where $|.|$ denotes cardinality and $J(A,x) \equiv \{r|\,\exists a \in A,\, a \prec r \prec x\}$ (Fig 5). The past of a maximal element m of $T_n(A)$ then casts a "shadow" on A, $J^-(m) \cap A$, and the set of shadows from each maximal element provides a covering $\mathcal{O}_v(A)$ of A. In [36] it was shown that the nerve simplicial complex $\mathcal{N}_v(A)$ constructed from $\mathcal{O}_v(A)$ gives the correct continuum homology for a large range of contiguous values of v, provided A satisfies certain conditions. Subsequent numerical work using ensembles of randomly selected inextendible antichains [37] suggests that these conditions are satisfied quite generically. This work lends support to a homological version of the fundamental conjecture, namely, that if a causal set faithfully embeds into two distinct globally hyperbolic spacetimes, then these are "approximately" homological.

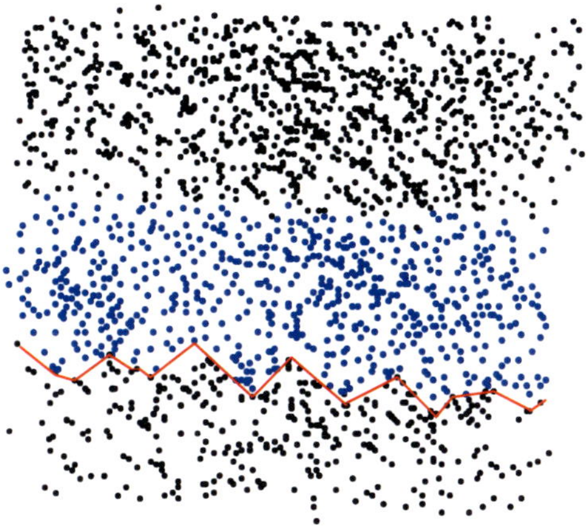

Figure 5. A diagram of a causal set that is approximated by a region in 2d Minkowski spacetime. The antichain in marked as a jagged red line and the blue region denotes an associated thickened antichain.

Although these results are robust, they nevertheless rely on a splitting of spacetime into space and time which is unnatural in a causal set context. It is still an open question whether interesting topological invariants can be obtained without recourse to inextendible antichains. Since Alexandroff intervals do not give rise to locally finite coverings of the causal set, is it possible to get around this, and work with non-locally finite coverings? Or are there other poset structures that one might use to define a new, more pertinent topology? Moreover, while homology is an important characterisation of topology, there are other topological invariants, some of which could be of importance to causal sets. Recent explorations on a Gauss-Bonnet-like relation in 2-d causal sets are suggestive and it would be interesting to gain a deeper understanding of these results [38].

The simplest geometric quantity that can be constructed in a causal set is the analog of timelike distance. For a generic pair $x \prec y$ in C, there are several chains that can be constructed with starting point x and end point y. Each chain can be thought of as a possible time-like curve from x to y. Since the continuum timelike distance is obtained by maximising over the length of timelike geodesics, a natural discrete analog of timelike distance is the length of the longest chain between x and y (Fig 2) [32]. For a causal set that is approximated by an interval in flat spacetime it was shown by Brightwell and Gregory [39] that this length converges rapidly to a multiple of the proper time τ, for large τ. This multiple is known for $d = 1, 2$ and has a fairly stringent bound for higher dimensions. Numerical simulations [40] suggest that the timelike distance estimator holds in 2,3 and 4 conformally flat spacetimes. More recently, progress has been made on recovering spacelike distances in flat spacetime [41]. Extending these results to generic curved spacetime are important but one will, again, have to address the question of localisation.

Using the antichain as a localisation device, we can also obtain the induced geometry of a spatial hypersurface Σ, with the volume parameter defined in the homology estimation being used to define a spatial metric. If A is an antichain which is approximated by Σ and $a_i, a_j \in A$, one can define the predistance function $\tilde{d}(a_i, a_j)$ as the smallest n such that $a_i, a_j \in J^-(x)$ and $|J(A, x)| = n$ if it exists, and infinite otherwise. This gives a labelled graph on A with vertices a_i and edges labelled by $\tilde{d}(a_i, a_j)$. The distance function $d(a_i, a_j)$ is then the shortest path on A. A similarly constructed distance function in the continuum then allows an analytical comparison with the continuum spatial geometry [42], but numerical simulations have not yet been carried out.

Apart from direct topological and geometric constructions, a causal set is also characterised by the way fields propagate on it. The causal set analog of a Green's function for a given field is thus an intrinsic property of that causal set. Every causal set C has a representation as a causal matrix M obtained from an order preserving labelling of C: M_{ij} is 1 if $i \prec j$ and 0 otherwise, where i, j denote labelled elements of C. Similarly, one can construct a link matrix L, for which L_{ij} is 1 or 0 depending on whether i is linked to j or not. In [43] it was shown that for a causal set which is approximated by 2d Minkowski spacetime, the Green's function for a scalar field satisfies $G_{ret} + G_{adv} = \frac{1}{2}(M + M^T)$ [31, 43] and a similar relation holds using the link matrix for causal sets that approximate to 4-d spacetimes. The D'Alembertian for this propagator is obtained by symmetrising G_{ret} and then inverting it. Simulations show that when restricted to causal sets that approximate to spacetime intervals, the continuum D'Alembertian can be recovered for suitable test functions.

A different approach was adopted in [31, 44] for constructing the D'Alembertian operator, which has been very successful both in resolving the problem of non-locality as well as admitting a generalisation to curved spacetimes [3, 4]. The basic insight used for taming non-locality comes from the following observation in [31]. Consider a field $\phi(u, v)$ in Minkowski spacetime that is slowly varying in a particular frame. If (u, v) are the lightcone coordinates in this frame, the D'Alembertian $\Box\phi(u, v) = \partial_u\partial_v\phi(u, v)$ on an interval $I(p, q)$ with $q = (u, v)$ and $p = (u - a, v - a)$ of area a^2 admits a discretisation $\Box_d = \phi(u, v)\frac{1}{a^2}(\phi(u, v) - \phi(u - a, v) - \phi(u, v - a) + \phi(u - a, v - a))$. Even though the system is Lorentz invariant, the contribution from other intervals of volume a^2 (corresponding to boosts of I) need not be the same since a given field configuration is not itself required to be Lorentz invariant. In particular, consider the contribution from a "narrow" interval $I'(p', q)$ with $p' = (u - a_1, v - a_2)$, where $a_1 a_2 = a^2$. As the interval becomes narrower, i.e., $a_1 \to 0$, the first two terms and the last two terms of the D'Alembertian each sum to small values because of the slow variation of the field. Thus, narrow intervals or those highly boosted with respect to the slowly varying frame do not contribute significantly to the D'Alembertian. Non-locality is embodied in a causal set by the existence of a large number of such narrow intervals of a fixed volume and this simple observation motivated the more complicated causal set construction which we now describe [31, 44].

In a causal set the nearest neighbours are links and hence one would expect an expression for the discrete D'Alembertian to include the sum $\sum_{y \in N_1^-(x)} \phi(y)$ where $N_1^-(x)$ denotes the set of elements that are in the past of x and linked to it. Since all these terms come with equal weight, one has to counter this with terms from the next nearest neighbours, and so on. In other words, the discrete D'Alembertian should be a sum of terms of alternating sign, with each term of the form $\sum_{y \in N_i^-(x)} \phi(y)$, where $N_i^-(x)$ denotes the set of elements to the past of x whose order interval $J(y,x)$ with x, contains precisely $i-1$ elements besides x and y. As with any approximation, it seems reasonable to truncate the operator by including only a finite number of layers and in order to obtain the correct continuum approximation, the exact coefficient with which each such "layer" appears in the sum has to be determined.

In 2d, for a volume cut-off l^2, the discrete operator takes on a particularly simple form [31]

$$B\phi(x) = \frac{4}{l^2}\left(-\frac{1}{2}\phi(x) + \sum_{y \in N_1^-(x)} \phi(y) - 2\sum_{y \in N_2^-(x)} \phi(y) + \sum_{y \in N_3^-(x)} \phi(y)\right), \quad (1)$$

namely, there is a truncation at the third layer. The mean of this expression over Poisson sprinklings of C converges to the continuum expression for the D'Alembertian as $l \to 0$. This averaging process also allows an expression in terms of an integral kernel

$$B(x-y) = \frac{4}{l^4} p(\zeta) e^{-\zeta} - \frac{2}{l^2}\delta^{(2)}(x-y) \quad (2)$$

where $p(\zeta) = 1 - 2\zeta + \frac{1}{2}\zeta^2$ and ζl^2 is the volume of the order interval $J(y,x)$ in units of l^2. However, although the expression (1) averages to the right continuum value, the fluctuations grow like N. In [31] a mesoscale $l_k >> l$ was introduced to damp these fluctuations and to rewrite the integral kernel (2) in terms of l_k. Discretising this kernel one gets the discrete operator

$$B_k\phi(x) = \frac{4\epsilon}{l^2}\left(-\frac{1}{2}\phi(x) + \epsilon \sum_{y \prec x} f((n(x,y),\epsilon))\phi(x)\right) \quad (3)$$

where $f(n,\epsilon) = (1-\epsilon)^n \left(1 - \frac{2\epsilon n}{1-\epsilon} + \frac{\epsilon^2 n(n-1)}{(1-\epsilon)^2}\right)$ and $\epsilon = l^2/l_k^2$. In writing this expression we have reincorporated terms from more than the first three layers but in the $l_k \to l$ limit one recovers the expression (1). The introduction of l_k means that rather than summing over order intervals of volumes in units of l^2

one is doing so over thickened layers with volumes in units of l_k^2. This effectively damps out fluctuations although it does lead to the appearance of a new non-locality scale. This form of the D'Alembertian has now been generalised to all spacetime dimensions [3, 4, 45] and a prescription for finding the exact coefficients is given in [4].

What is the generalisation of this construction to curved spacetime? It turns out surprisingly enough that one can indeed use the very same flat spacetime expression for the operator B_k, except that additional curvature dependent terms also arise [3]. Using the Gibbons-Soludukin expression for the Ricci scalar in terms of the volume and height (or proper distance) of a small causal intervals $J(x, Y)$ [33] the authors of [3] showed that

$$\lim_{l_k \to 0} B_k \phi(x) = \left(\Box - \frac{1}{2} R(x) \right) \phi(x) \tag{4}$$

in both 2 and 4 spacetime dimensions, where ϕ is assumed to vary slowly with respect to l_k and the radius of curvature is much larger than l_k. Using the constant field, this was then used to obtain a causal set version of the Einstein-Hilbert action, the Benincasa-Dowker action in 2 and 4 dimensions

$$\frac{1}{\hbar} S^{(2)}[C] = 4(N - 2N_1 + 4N_2 - 2N_3) \tag{5}$$

$$\frac{1}{\hbar} S^{(4)}[C] = 4(N - N_1 + 9N_2 - 16N_3 + 8N_4) \tag{6}$$

where N is the number of elements in the causal set C and N_i the number of elements in the order interval which contains precisely $i - 1$ elements besides the end points. Thus, N_1 counts the number of links or 2 element chains, N_2 the number of 3-element chains, N_3 the number of 4-element chains plus the number of diamond causal sets. The Benincasa-Dowker action is a clear proof that causal sets can give rise to local physics, and is a big milestone for the theory. The action was generalised to arbitrary dimensions in [4].

Earlier work on finding a causal set version of the Einstein-Hilbert and matter actions for causal sets adopted a somewhat different approach [46]. The Gibbons-Soludukin results were used to find certain components of the Ricci tensor in terms of the volume and heights of small spacetime intervals, and in order to find a way of including all possible intervals, a minimisation procedure was adopted. However, this procedure does not lead to a closed form for the action. It is possible that a procedure similar to the layer construction of [31] can be used and this could be a useful direction to pursue.

One of the promises one tends to associate with discretisation is that of curing the UV divergences of quantum field theory. Typical discretisations break Lorentz invariance since a UV cut-off is put in by hand. As we have seen, causal set discretisation is very different in that it preserves Lorentz invariance and it is therefore an open question whether it can tame these divergences. A reasonable hope might be that causal set discretisation can provide a Lorentz invariant renormalisation procedure but one is currently a long way from showing this [31]. To begin with, one has to first construct a quantum field theory on a causal set which is approximated by Minkowski spacetime. In [5] a Feynman-diagram inspired prescription was given to obtain the retarded Green's function $G(x-y)$ of a quantum scalar field in 2 and 3 spacetime dimensions. In 2-d, one sums over all chains from x to y, weighting each vertex or element with an amplitude b and each leg or relation that appears in the chain by an amplitude a. In 4-d one sums only over chains in which all the relations are strictly links. Comparison of the continuum propagator with the discrete version allows the amplitudes a and b to be evaluated. The 2d causal set propagator is well approximated by the continuum and in 4d, this is true in the continuum limit.

In [6] the next big step was taken by constructing the Feynman propagator using a rather novel procedure. To begin with, the Pauli-Jordan function is a matrix on the causal set $\Delta = G_R - G_A$, where G_R and G_A are the 2d or 4d causal set versions of the retarded and advanced Green's functions that were obtained previously. $i\Delta$ is a Hermetian skew-symmetric matrix, and hence of even rank $2s$. This means that its non-zero eigenvalues appear in real positive and negative pairs $(\lambda_i, -\lambda_i)$ with $i = 1, \ldots, s$. The associated positive and negative eigenvectors $\{u_i\}$ and $\{v_i\}$ are used to define a matrix $Q = \sum_{i=1}^{s} \lambda_i u_i u_i^\dagger$, so that $i\Delta = Q - Q^*$, and bosonic quantum field operators $\hat{\phi}_x$ are associated with each causal set element. Apart from satisfying the standard Hermiticity and commutation conditions these operators satisfy a new condition which replaces the standard requirement that $\hat{\phi}$ satisfy the Klein Gordon equation. Namely, all operators linear in $\hat{\phi}$ which commute with $\hat{\phi}$ are required to themselves be zero. Thus

$$i\Delta w = 0 \Rightarrow \sum_{x=1}^{N} w_x \hat{\phi}_x = 0, \tag{7}$$

where N is the number of causal set elements. The positive and negative eigenfunctions $\{u_i\}$ and $\{v_i\}$ of $i\Delta$ allows one to define creation and annihilation operators and using the conditions on $\hat{\phi}_x$, a consistent mode decomposition for the field operator is obtained. The vacuum associated with the creation and annihilation operators is then used to construct the two point function $Q_{xy} = \langle 0|\hat{\phi}_x\hat{\phi}_y|0\rangle$ and the natural order-preserving labelling on the causal set gives the requisite time-ordering. The Feynman operator is then given by $G_F = G_R + iQ$. Numerical tests in 2-d and 4-d demonstrate that the discrete operator has the recognisable continuum form as the discreteness scale goes to zero.

What is makes this construction particularly interesting is that a vacuum state is obtained purely algebraically without explicit reference to a coordinate system. The positive and negative mode decomposition arise from properties of the Pauli-Jordan matrix which itself is defined independent of any frame of reference. Whether this construction has consequences for quantum field theory on curved spacetime is clearly an important direction to pursue. What special role does the new ingredient, Eqn (7) play, for example?

The existing calculations for the blackhole area and entropy bounds can also be categorised as kinematics since the causal set itself remains non-dynamical [47, 48, 49]. In [47] the area of a black hole like region is calculated by counting the links that cross the horizon. This and similar calculations require a past volume cut-off, but yield the desired geometric result, namely that the area of the blackhole horizon is proportional to the number of links that cross it. Whether assigning these links the term "horizon degrees of freedom" offers a glimpse into a more complete description of this relationship can only be determined by a better understanding of the quantum dynamics of causal sets.

3. Phenomenology

An important criticism levied against quantum gravity by most other physicists is that it offers us no experimental signatures. Almost as a response to this criticism, one has come across a virtual plethora of models of quantum gravity in which one sees the introduction of large extra dimensions and a subsequent reduction of the scale of quantum gravity on the one hand, and violations of the most stringently verified symmetry of nature, Lorentz invariance. It is therefore of interest to see what it is that causal set theory can offer by way of an experimental signature. Without a complete theory of causal set quantum gravity – kinematics *and* dynamics – one may worry that any model building is futile. However, there is a big desert of energy scales between the Plank energy scale and that of current colliders and it is a legitimate query to ask whether, given a causal set discretisation of a spacetime, there are effects of the discreteness scale on observation. The small effects of discreteness can be amplified by cosmological distances and hence there are potential cosmological signatures that need to be explored.

The most significant cosmological signature from causal set theory is the prediction for the value of the cosmological constant in the late '80s [28]. This preceded the actual experimental observation of a small but non-zero value for Λ, which, strikingly enough, is of the same order of magnitude as the causal set prediction. The basic argument is very general and is a consequence of any theory which assumes a random discreteness for spacetime and satisfies the number to volume correspondence (2). The cosmological constant appears in the Einstein-Hilbert action as the volume term $\Lambda \int \sqrt{-g} d^4x = \Lambda V$ where V can be taken to be the total past volume of the universe. Fixing this volume in a unimodular modification of gravity gives rise to an uncertainty relation between the fluctuations in Λ and Δ: $\Delta\Lambda\Delta V \sim 1$ in natural units. In a causal set discretisation of the FRW universe, the fluctuation in volume can be determined from the Poisson distribution, i.e., $\Delta N = \sqrt{N}$ so that $\Delta V = \sqrt{V}$. One finds that $\rho_\Lambda \sim \sqrt{V}^{-1} \sim H^2 = \rho_{critical}$ at all times and that given the age of our universe and hence V, $\Delta\Lambda$ in our present epoch is the small value 10^{-120} in natural units. Such a small value for Λ has been notoriously difficult to calculate from standard quantum field theory arguments and hence it is impressive that it emerges so naturally from causal set theory. On the other hand, it is only the *fluctuation* that causal set theory can currently predict, and this matches the observation only when the mean value of Λ is assumed to be zero.

In [50] a stochastic model for the evolution of a fluctuating Λ was incorporated into a modified FRW model which retains only the Hamiltonian constraint. The stochastic models are characterised by a single free parameter α and for suitable choices of α numerical simulations show a Λ which fluctuates between negative and positive values with a significant fraction of the trials giving rise to a small but positive cosmological constant in the present epoch. In these same trials, the Big Bang Nucleosynthesis phase is characterised by a large negative Λ which the authors suggest may help explain the reduced Helium abundance of observations. Expanding this stochastic model to one in which Λ varies spatially is of great importance in getting observational constraints on the free parameters [51]. Moreover, an answer to why the mean value of Λ is zero may only be possible in a more developed cosmological model which incorporates quantum fields.

As we have noted, the basic argument for a fluctuating Λ is very simple and seemingly transcends the details of the theory. Indeed, it has been suggested that since spacetime discreteness is a standard feature of several approaches to quantum gravity, they must also admit a similar solution to the cosmological constant problem. This would in principle be reasonable if we could view this problem purely as a macroscopic manifestation of a fundamental theory. After all, several disparate approaches to quantum gravity yield the same black hole area law. However, it is the precise manifestation of causal set discreteness via a number to *spacetime* volume correspondence that distinguishes it from the other approaches and is key to the arguments of [28]. If this correspondence is taken seriously, then one is forced to consider a random discretisation of spacetime and no other approach to quantum gravity currently uses this as a starting point.

Are there other observational consequences of causal set discreteness? The Poisson sprinkling of causal set elements allow for the possibility of very large voids or continuum volumes which are entirely unpopulated by causal set elements. Such voids, if ubiquitous, would surely contradict experimental observation since in such regions the continuum description would entirely breakdown. In spacetimes of infinite extent it is indeed the case that there will almost surely be arbitrarily large voids, but it is more pertinent to ask this question of our observable universe. In [52] it was argued that the probability of a single nuclear sized void in the observable universe is less than $\sim 10^{252} \times e^{-10^{72}}$! Thus, one needs to find more significant observational signatures for causal set discreteness.

In [52] a model of a massless classical particle propagating on a causal set was constructed. A causal set C which faithfully embeds into flat spacetime offers no straight line path for a particle which thus must "swerve", i.e., the direction of the momentum must change while hopping from one element e_n to the next e_{n+1}. The momentum p_{n+1} associated with such a hop can be taken to be proportional to the spacetime vector in the spacetime embedding, with starting point e_n and end point e_{n+1}. Since this is a classical motion, the particle is assumed to choose a path in C which is as close to a straight-line as possible. This means that, given a particular pair (e_n, p_n), the next element e_{n+1} is chosen as close to the direction p_n as possible so that $|p_{n+1} - p_n|$ is minimised. Additionally, e_{n+1} is chosen to lie within a certain proper time τ_f which corresponds to a "forgetting time". In the continuum this process manifests itself as a proper time diffusion in phase space $\mathbb{H}^3 \times \mathbb{M}^4$, where $\mathbb{H}^3$ is the momentum mass shell and $\mathbb{M}^4$ is Minkowski spacetime. The probability distribution $\rho(p^\nu, x^\mu, \tau)$ then satisfies the Lorentz invariant diffusion equation [52, 53]

$$\frac{\partial}{\partial \tau}\rho = k\Delta_p^2 \rho - \frac{1}{mc^2} p^\mu \frac{\partial}{\partial x^\mu}\rho, \tag{1}$$

where k is a parameter which will depend on the forgetting time τ_f. Assuming the probability distribution to be spatially uniform, the Gaussian evolution of the probability in momentum space means high energy events become significant over sufficiently long periods of time. This affords a Fermi-acceleration type mechanism for generating very high energy cosmic rays. For protons, laboratory and astrophysical data can be used to put too stringent a limit on k for protons [54]. Nevertheless, the possibility of neutrino sources giving rise to similar signatures hasn't yet been ruled out. Since the mechanism does not depend on a specific matter distribution, if there are causal set origins for ultra high gamma rays then these should be distributed isotropically. Moreover, as in a standard diffusion process the swerves do not conserve energy, but as discussed in [55] this does not contradict the current astrophysical bounds on Lorentz violation. Subsequent work in this direction includes the massive particle models of [56, 57].

More recently, a model has been proposed to study the effects of Lorentz invariant discretisations on the CMB polarisation [58]. Here, in addition to the phase space one has the set of polarisation states $\mathcal{B}$ given by the Bloch sphere. Neglecting effects of phase spacetime diffusion, one gets a Lorentz invariant diffusion on $\mathcal{B}$ which reduces for linearly polarised states to a diffusion cum a drift equation. In the cosmic frame these are frequency dependent. The authors of [58] propose that this can give rise to signatures in CMB polarisation and the hope is that this exciting proposal will find useful constraints when new data becomes available from Planck.

Before going on to examining dynamics, it is perhaps important to emphasise that while it may seem tempting to some to construct a $3+1$ Cauchy-type formulation of causal sets based on thickened antichains, such an attempt is doomed to failure for a reason that can, once again, be traced to non-locality. Instead of being able to mimic a Cauchy surface whose intrinsic and extrinsic geometry captures all the required information for future evolution, an inextendible antichain A is like a "sieve"– a large amount of geometric information in fact by-passes it, because of what I will call "missed links". Consider a causal set C that faithfully embeds into Minkowski spacetime and let A be an inextendible antichain in C. While $J^+(A) \cap J^-(A) = A$, where $J^+(A) \equiv \{r | \exists\, a \in A,\, a \prec r\}$ and similarly $J^-(A) \equiv \{r | \exists\, a \in A,\, r \prec a\}$, so that A neatly divides the *elements* of C into the disjoint subsets $C = A \sqcup (J^+(A) \backslash A) \sqcup (J^-(A) \backslash A)$, this is not true of the relations. In particular there exists pairs p, q with $p \in J^-(A) \backslash A$ and $q \in J^+(A) \backslash A$ such that $p \prec q$ is a link, which A however "misses" (see Fig 6). In fact A misses an infinite

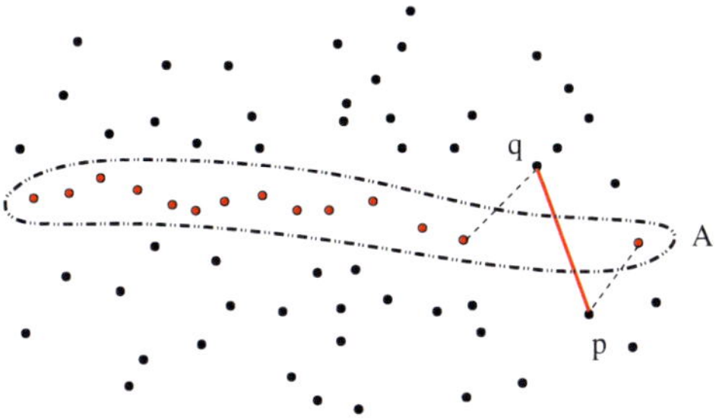

Figure 6. A missing link $p \prec q$ through an inextendible antichain A is marked with a bold red line.

number of links in Minkowski spacetime. Are these causal links relevant to signal propagation [30, 52, 56, 59]? If so, then far from capturing all the information relevant to its future, A in fact lets most of it slip through! A finite volume thickening of A clearly cannot eliminate this problem, and one has to contend with the question of how to recover the approximate global hyperbolicity of our physical laws. It is important that the missed links tend to be those that are highly boosted with respect to the rest frame associated with the antichain and hence it seems physically reasonable that such a signals will tend to pass through a macroscopic body unnoticed. However, given that there are an infinity of such links in Minkowski spacetime it is important to know whether the effect is cumulative. Does it imply, for example that in order for causal sets to be compatible with known physics, that the universe be past finite, at least until the last cosmological bounce?

4. Dynamics

Given that causal sets do not admit a natural space and time split, it is clear that a Hamiltonian framework is not viable for describing causal set dynamics. Indeed, it is far more natural to talk of dynamics in a path integral or sum-over histories language. In a path integral representation of 4-dimensional quantum gravity, the histories space is the space of all Lorentzian 4-geometries. In causal set theory this is replaced by the set of all (unlabelled) locally finite posets Ω. Importantly, Ω includes causal sets that do not have a continuum approximation, and the hope is that a suitable dynamics would, in the classical limit, pick out causal sets that are approximated by continuum spacetimes. But does a typical causal set look anything like a spacetime and if not does it only differ from it "in the small"? Dynamical triangulations have already taught us that such an expectation is naive, however reasonable the discretisation may seem – it is a well know problem that most simplicial manifolds have a branched polymer structure, which dominate the path integral [60]. Causal sets also potentially face such an entropy problem. As the size of the causal set N gets larger, most causal sets are of the Kleitman-Rothschild or (KR) form shown in Fig 7 [61]. These have only three layers, with a large number of relations between each layer and are therefore most unlike spacetime. However, their numbers grow like $2^{N^2/4}$ where N is the size of the causal set and hence seem like a threatening presence in the path integral. Moreover, it was shown in [62] that apart from the KR posets there is a hierarchy of classes of sub-dominant causal sets, with the number of layers increasing as one goes down the hierarchy. The entropic landscape is therefore complex and characterised by first order transitions between these sub-dominant classes [62]. Any causal set dynamics must therefore work hard to counter this dense entropy landscape.

It is thus an important and surprising result that a dynamics can be constructed which counters this entropy. These are the Rideout-Sorkin classical sequential growth models alluded to in the abstract [2, 63]. These models use a

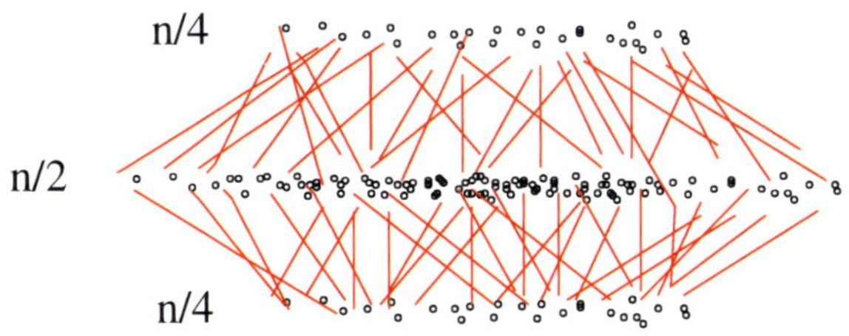

Figure 7. A three layer n element Kleitman-Rothschild poset.

bottom-up approach to dynamics – rather than deduce a rule from the continuum one begins with a set of fundamental principles. Starting from the empty set, a causal set is grown element by element in a Markovian fashion by assigning probabilities to each transition from an n element causal set to an $n+1$ element causal set. The $n+1$th element can be added either to the future of or be unrelated to an existing element, but it cannot be added to its past. Apart from being Markovian, the dynamics is required to obey label invariance (a discrete avatar of covariance) as well as a causality condition dubbed Bell causality. The latter encodes the requirement that the transition probabilities should be independent of the "spectator" elements, i.e., those that are not involved explicitly in the transition. These simple rules can be manipulated to extract a set of coupling constants, t_n, one for each stage of the growth, which then fully determine the dynamics. In the cosmological context, starting from generic initial conditions, these coupling constants are seen to get renormalised after every cosmological epoch separated by a "bounce", with a line of fixed points corresponding to a simple, transitive percolation dynamics, with $t_n = t^n$ for $t \geq 0$ [64]. This last era of transitive percolation gives rise typically to causal sets which have more than a fleeting resemblance to continuum spacetime [65] although they do not seem to admit an exact spacetime approximation. Nevertheless they teach us that it is possible to overcome the threat of KR posets with an appropriate choice of dynamics.

Since the growth process cannot add a new element to the past of an existing one, the stochastic growth models give rise to causal sets that are past finite. Each causal set is moreover labelled by the specific growth process and hence, even though the probabilities are label invariant, the causal sets themselves are labelled. Because of the nature of the growth the labelling is order preserving, i.e., $x \prec y \Rightarrow l(x) < l(y)$ where $l(x) \in \mathbb{N}$ denotes the labelling of the causal set element x. Thus, the space of histories or sample space Ω is the space of past finite labelled causal sets. An obvious and important question is – what are the covariant or label invariant questions that one can ask of such a system? This question was addressed in [66] by re-expressing the dynamics as a probability measure space. A finite n-element labelled causal set c_n can be used to construct a cylinder subset $\mathrm{cyl}(c_n)$ of Ω which is the set of all labelled causal sets in Ω whose first n elements are the causal set c_n. The event sigma algebra $\mathcal{S}$ is then constructed by taking countable unions and intersections of all such cylinder sets constructed from the set of finite labelled causal sets. In the growth process, any c_n is assigned a probability $\mathcal{P}(c_n)$ and hence we can take $\mathcal{P}(\mathrm{cyl}(c_n)) \equiv \mathcal{P}(c_n)$. This measure is known to extend to all of $\mathcal{S}$ by the Kolmogorov-Caratheodary-Hahn extension theorem, so that the triple $(\Omega, \mathcal{S}, \mathcal{P})$ is a well defined probability measure space. A label invariant measure space can then be constructed formally [66] by taking its quotient under relabellings: if $\widetilde{\Omega}$ is the set of unlabelled past finite causal sets, then $\widetilde{\mathcal{S}}$ is the sub-sigma algebra of $\mathcal{S}$ constructed as follows: a subset $\tilde{S} \subset \widetilde{\Omega}$ is an element of $\widetilde{\mathcal{S}}$ if for every $c \in \tilde{S}$, its relabelling c' is also in $\tilde{S}$. The restriction of $\mathcal{P}$ to $\widetilde{\mathcal{S}}$ then provides a label invariant measure space $(\widetilde{\Omega}, \widetilde{\mathcal{S}}, \mathcal{P}_{\widetilde{\mathcal{S}}})$ [66].

But what are these label invariant or covariant events and can they be assigned a physically useful meaning? Of interest to quantum cosmology are covariant questions of the type: how many bounces has the universe gone through? In [66] it was shown that the formal event algebra $\widetilde{\mathcal{S}}$ can be replaced, upto set of measure zero, with a sub-sigma algebra $\mathcal{S}'$, which is generated from physically well defined events. A finite unlabelled sub-causal set $\tilde{c_n}$ of $\tilde{c} \in \widetilde{\Omega}$ is said to be a partial stem if it contains its own past, and is the causal set analogue of a "past-set" $J^-(X)$ in a spacetime. A stem set $\mathrm{stem}(\tilde{c_n})$ is a subset of $\widetilde{\Omega}$ such that every $\tilde{c} \in \mathrm{stem}(\tilde{c_n})$ contains the partial stem $\tilde{c_n}$. Thus, for these cosmologies it is possible to ask, for example, what the probability is for the universe to have a single element which lies to the past of every other element (originary growth). For transitive percolation, this is given by the Euler function $\phi(q) = \Pi_{i=1}^{\infty}(1 - q^i)$, where $q = 1 - p$ is the probability for a new element to be unrelated to all other elements [67]. Thus, the importance of this sub-sigma algebra is clear – it makes it possible to pose covariant cosmological questions and to therefore construct the covariant "observables" or events relevant to cosmology. It is the hope that at least a part of this analysis survives into a quantum dynamics for causal sets.

These stochastic growth models in fact admit a simple quantum generalisation with the transition probabilities replaced by transition amplitudes. Instead of a probability measure space the growth process generates a quantum pre-measure space $(\Omega, \mathcal{A}, \mu)$, where $\mathcal{A}$ is the algebra generated from cylinder sets, closed only under finite unions and intersections, and μ is a quantum pre-measure $\mu : \mathcal{A} \rightarrow \mathbb{R}^+$ which satisfies the more general quantum sum rule (QSR) [68, 69, 70]:

$$\mu(A \cup B \cup C) = \mu(A \cup B) + \mu(A \cup C) + \mu(B \cup C) - \mu(A) - \mu(B) - \mu(C), \quad (1)$$

for the disjoint sets $A, B, C \in \mathcal{A}$. μ is not in general additive since there can be interference between different sets in $\mathcal{A}$. This means that the extension of μ to the full event sigma algebra $\mathcal{S}$ is not guaranteed. But such an extension is required if we want to construct covariant observables in analogy with the classical growth models. Using the Histories Hilbert space construction of [71] the quantum pre-measure can be recast as a vector pre-measure [67] for which a Kolmogorov-Caratheodary-Hanh type extension theorem exists provided μ satisfies certain condition. This analysis was used to explore certain simple "complex percolation" models in [67]. These models are useful quantum generalisations of the classical stochastic theories since they incorporate both the Markovian evolution as well as label invariance[4].

Underlying the above construction of a quantum measure space is another deep current of enquiry into the measurement question of quantum theory. The interpretational problems of quantum theory become of tantamount importance in quantum cosmology where one needs to give meaning to cosmological events in the absence of measuring devices. A quiet revolution in quantum interpretation was initiated by Sorkin in a series of papers starting from a quantum measure formulation [68, 69] and culminating more recently in the anhomomorphic logic proposal [74]. This is a realistic interpretation of quantum theory, independent of external observers and measuring devices, which nevertheless passes the stringent Kochen-Specker test [75]. While an exciting development for quantum theory, a more detailed description is outside the scope of this review. From

[4]However, is not quite clear what the proper quantum analog of the Bell causality condition is [72, 73]. While its easy to implement a classical analog of this condition in these simple models, a more general implementation seems riddled with complications.

a functional point of view however, it suffices to think of causal set dynamics simply as a quantum measure space. Questions related to observation reduce to events that are measurable – an event of quantum measure zero, for example, can be taken to "not occur".

The construction of the causal set action described in Section 2 on the other hand allows a "top-down" approach to quantisation along more conventional lines (keeping the interpretational issues at bay). Using a parameter Wick rotation of the type described in [76] the Lorentzian action can be Euclideanised to obtain a partition function for causal sets. Importantly, unlike the standard Euclideanisation which changes the set of histories from Lorentzian to Euclidean geometries, this procedure does not change the space of causal sets[5]. Causal set dynamics can then be simulated using Monte Carlo methods and efforts in this direction are currently underway [77]. Early simulations show promise, though much work remains to be done in understanding the complex phase structure of the entropy landscape. While an important direction to pursue, however, from a fundamental point of view it is unclear that the causal set action *should* provide us with the correct quantum dynamics. It is entirely possible that dynamics is dictated by yet undiscovered fundamental rules and that the Einstein-Hilbert action makes its appearance only as an effective dynamics much like the Navier-Stokes equation.

An interesting, though limited, example of a top-down approach to constructing causal set quantum is a 2d model of causal sets which exhibits unexpected results [78]. It is important to reiterate that unlike most other theories of quantum gravity in which dimension is inbuilt, Ω includes causal sets that approximate to spacetimes of *all* dimensions. Hence much more is expected from causal set dynamics than from other approaches, since spacetime dimension is also required to be a prediction of the theory. On the other hand it is a useful exercise to examine such a dimensional restriction of causal sets. All finite element causal sets that are approximated by conformally flat 2-d spacetimes with trivial topology belong to the class of 2d-orders, obtained by taking the intersection of two linear or total orders. One can think of these two total orders as the discrete lightcone coordinates (u, v) of a causal set that faithfully embeds into such a spacetime – an element $e_1 = (u_1, v_1)$ precedes an element

[5]Indeed, causal sets are so inherently Lorentzian it is difficult or impossible to imagine a Euclideanised version!

$e_2 = (u_2, v_2)$ iff both $u_1 < u_2$ and $v_1 < v_2$. Note that in a sprinkling, the probability of an element to lie exactly on the light cone of another element (i.e., either their u or v coordinates are the same) is precisely zero. However, not all 2d orders admit a continuum approximation, and thus, though there is an intrinsic causal set dimension given by the number of total orders, this does not itself imply manifoldlikeness. In particular it means that the entropy problem could also potentially rear its head in such a model. What is surprising is that even for a uniform measure on 2d orders, the dominant contribution comes from those which faithfully embed into *flat* 2d spacetime [79, 80, 78]. A uniform measure can be motivated from a continuum inspired model of quantum gravity [78] which soups up this order theoretic result into a genuine 2d model of quantum gravity. Generalisations to higher dimensions though much harder to do, would be a natural direction to pursue.

5. Conclusion

What is the current state of causal set theory? For years, criticisms have ranged from saying, on the one hand, that the approach is too minimalist to give us any recognisable physics, and on the other that it is too ad hoc and not well motivated enough. The latter complaint in my opinion is entirely unjustified. *All* approaches to quantum gravity necessarily use new ingredients, even when guised in the modest robes of "naturalness". Questions of choice and emphasis are motivated either by principles of simplicity (however that is interpreted) and by phenomenological restrictions. Causal set theory is well motivated in this sense – a causal set is the simplest possible manifestation of a quantum *Lorentzian* geometry. It takes the Lorentzian character of spacetime, in particular the *classical* results of [13, 16], most seriously adding only the simplest ingredient of discreteness in a manner *consistent* with the classical results. They are also, at a basic level, phenomenologically justified since they satisfy Lorentz invariance, a well verified symmetry of nature. The fact that it is too minimalist is a criticism that perhaps understandably kept interested researchers at bay – how could one do any realistic physics at all with a construction which gives rise to such apparently uncontrollable non-locality? In retrospect, these fears are unfounded given the recent developments in the field, in particular the construction of a well defined local D'Alembertian operator and Einstein-Hilbert action for causal sets. It gives hope that with continued work other seemingly insurmountable hurdles could also be one day be cleared.

I now present a short list of questions of immediate interest to causal set workers. It is far from exhaustive, but hopefully it is representative. Arguably, one of the most interesting questions is whether the formalism developed in [5, 6] can be applied to black-holes to study Hawking radiation. Discreteness is often taken to be synonymous with a ultraviolet cut-off, but this is not true in causal set theory, since the discretisation is Lorentz invariant. It is therefore not clear whether causal sets cure the transplankian problem. Is there instead a manifestation of non-locality which modifies the thermal spectrum as in other non-local theories [81]? Constructing a decoherence functional (or quantum measure) in terms of the Feynman propagator is also an important question to address, as is the question of how interactions manifest themselves in this framework. In [82] an interesting proposal has been made for constructing an S matrix on the causal set for interacting fields and it is important to know how far this analysis can be taken. In [31, 3, 4] in the process of constructing discrete operators, non-local continuum actions were constructed and it would be an interesting exercise to study such non-local continuum models and obtain constraints on the non-locality scale.

The numerical work on the Monte Carlo simulations for quantum dynamics of causal sets, while still in its infancy, should be able to tell us whether a Euclideanised dynamics of causal sets can give rise (after a reverse Wick rotation) to causal sets that are manifoldlike in the classical limit. In the Causal Dynamical Triangulation(CDT) approach, a Euclideanised dynamics has been employed successfully to cure the entropy problem encountered when the causality constraint is not included [20] and it is a somewhat analogous success that we seek with causal sets. On the other hand, while interesting and of value, the success or failure of this approach may not be related, in a simple way, to the construction of a truly bottom-up quantum dynamics. As described in Section 4 one would like to obtain a fundamental class of quantum dynamics based on first principles much like the classical stochastic growth dynamics, but the implementation of causality remains an important open question. In addition, any examination of quantum dynamics ties in intimately with the other aspect of quantum gravity (which we have touched upon perhaps too lightly in this review), namely, that of the measurement problem. The quantum measure approach provides a framework to do so and while one now has an understanding of certain finite dimensional situations, developments in the infinite dimensional case will be crucial to quantum gravity [67]. Another very important question which, when answered, should make the approach more amenable to the wider community, is how matter should be incorporated into causal set theory. The combination of discreteness with Lorentz invariance means that there is no ready prescription for putting vector fields or higher spin fields on the causal set. Should these instead emerge from the theory or does one simply need an (as yet undiscovered) well defined way of putting them in "by hand"? Fermions are perhaps easier to incorporate – one can assign spins to the links, but relating these to spacetime fermions brings us back to the problem encountered with putting higher spins on the causal set. Progress on such questions would be extremely valuable in constructing more realistic phenomenological models. In [82] a suggestion for a Feynman chequerboard model was made for introducing spin on the causal set. These discussions suggest that rather than getting classical fields to sit on a causal set, one should construct the more fundamental quantum field operators which in the continuum approximation will give rise to the classical fields.

Another very important direction that needs to be actively pursued is the search for phenomenological signatures of the sort described in Section 3 To begin with, more realistic models of a varying Λ need to be constructed to match observations more precisely. Can the neutrino origins of cosmic rays give hope to the swerve model of particles and are there other ways in which the Lorentz invariant discreteness of causal sets can manifest itself in observations? One can at this stage ask bolder questions – can causal set theory explain the other big puzzles of cosmology like dark matter or structure formation, and can it provide an alternative explanations to the scale invariance of the CMB power spectrum? These may seem to broad a class of questions, but it seems possible at least in principle to build models which constrain the effects of spacetime discreteness on observations.

In this article I have outlined causal set theory in broad strokes and pointed to the directions which are currently being pursued. A modest hope is that the reader will become familiar with the basic tenets of this approach, while a more ambitious hope is that she or he will then begin actively ruminating on causal sets!

Acknowledgments

I am especially thankful to Rafael Sorkin, Fay Dowker, David Rideout and Joe Henson for valuable discussions over the years. This work was supported in part by the Royal Society International Joint Project 2006-R2. I would also like to thank the Physics and Computer Science departments at McGill University for their hospitality while this work was being completed.

References

[1] L. Bombelli, J. Lee, D. Meyer and R. D. Sorkin, *Phys. Rev. Lett.* 59 521-524 (1987).

[2] D. P. Rideout and R. D. Sorkin, *Phys. Rev. D* 61 024002 (2000).

[3] D. M. T. Benincasa and F. Dowker, *Phys. Rev. Lett.* 104 181301 (2010).

[4] H. F. Dowker and L. Glaser, In preparation.

[5] S. Johnston, *Class. Quant. Grav.* 25 202001 (2008).

[6] S. Johnston and S. Johnston, *Phys. Rev. Lett.* 103 180401 (2009).

[7] Roger Penrose, "Techniques of Differential Topology in Relativity", AMS Colloquium Publications (SIAM, Philadelphia, 1972).

[8] S. W. Hawking and G. F. R. Ellis, *The large scale structure of space-time*, Cambridge: Cambridge University Press, 1973.

[9] O. Parikkar and S. Surya, arXiv:1102.0936 [gr-qc].

[10] E. H. Kronheimer and R. Penrose, *Proc. Cambridge Phil. Soc.* 63 481 (1967).

[11] E.C. Zeeman, *J. Math. Phys.* 5 490-493 (1964).

[12] R. P. Geroch, E. H. Kronheimer and R. Penrose, *Proc. Roy. Soc. Lond. A* 327 545 (1972).

[13] S. W. Hawking, A. R. King and P. J. Mccarthy, *J. Math. Phys.* 17 174 (1976).

[14] R. P. Stanley, Enumerative Combinatorics, vol. 1, Wadsworth and Brooks/Cole, Pacific Grove, CA, 1986 G/A second printing, Cambridge University Press, Cambridge, 1997.

[15] D. Scott. Outline of a mathematical theory of computation. *Technical Monograph PRG-2,* Oxford University Computing Laboratory, November 1970.

[16] D. B. Malament, *J. Math. Phys.* 18 1399-1404 (1977).

[17] A.V. Levichev; *Soviet Math. Dokl.* 35 452-455 (1987).

[18] R. D. Sorkin, *Int. J. Theor. Phys.* 36 2759 (1997).

[19] R. D. Sorkin, Causal Sets: Discrete Gravity (Notes for the Valdivia Summer School), In *Proceedings of the Valdivia Summer School,* edited by A. Gomberoff and D. Marolf.

[20] J. Ambjorn, J. Jurkiewicz and R. Loll, *Phys. Rev. Lett.* 93 131301 (2004).

[21] N. H. Christ, R. Friedberg and T. D. Lee *Nucl. Phys. B*202 89 -125 (1982); T D Lee *Discrete mechanics*, in A Zichichi, ed How Far Are We from the Gauge Forces (XXI Ettore Majorana School, Erice 1983) Plenum.

[22] L. Bombelli, J. Henson and R. D. Sorkin, *Mod. Phys. Lett. A* 24 2579-2587 (2009).

[23] D. A. Meyer, *The Dimension of Causal Sets*, PhD thesis (M.I.T., 1988).

[24] L. Bombelli, Talk at "Causet 2006: A Topical School funded by the European Network on Random Geometry", Imperial College, London, U.K., September 18-22, 2006.

[25] L. Bombelli and D. A. Meyer, *Phys. Lett. A* 141 226-228 (1989).

[26] L. Bombelli, *J. Math. Phys.* 41 6944-6958 (2000).

[27] J. Noldus, *Class. Quant. Grav.* 21 839-850 (2004), L. Bombelli and J. Noldus, *Class. Quant. Grav.* 21 4429 (2004), arXiv:gr-qc/0402049.

[28] R. D. Sorkin, paper presented to the conference on The History of Modern Gauge Theories, held Logan, Utah, July 1987; *Int. J. Theor. Phys.* 33 523-534 (1994).

[29] T. Regge, *Nuovo Cim.* 19 558 (1961).

[30] Rafael D. Sorkin, *J. Phys.: Conf. Ser.* 174 012018 (2009).

[31] R. D. Sorkin, Does Locality Fail at Intermediate Length-Scales; *Towards Quantum Gravity,* Daniele Oriti (ed.) (Cambridge University Press, 2007);

[32] J. Myrheim, CERN preprint TH-2538 (1978).

[33] G. W. Gibbons and S. N. Solodukhin, *Phys. Lett. B* 652 103 (2007).

[34] D. D. Reid, *Phys.Rev. D* 67 024034 (2003).

[35] R. D. Sorkin, *Int. J. Theor. Phys.* 30 923 (1991).

[36] S. Major, D. P. Rideout and S. Surya, *J.Math.Phys.* 48 032501 (2007).

[37] S. Major, D. P. Rideout, S. Surya, S. Major, D. Rideout and S. Surya, *Class. Quant. Grav.* 26 175008 (2009).

[38] D. Benincasa and F. Dowker, In preparation.

[39] G. Brightwell and R. Gregory, *Phys. Rev. Lett.* 66 260-263 (1991).

[40] R. Ilie, G. B. Thompson and D. D. Reid, *Class. Quant. Grav.* 23 3275 (2006).

[41] D. Rideout and P. Wallden, *Class. Quant. Grav.* 26 155013 (2009).

[42] J. Henson, S. Major, D. P. Rideout and S. Surya, Work in progress.

[43] A. R. Daughton, "The Recovery Of Locality For Casual Sets And Related Topics," Ph.D. Thesis.

[44] J. Henson, arXiv:gr-qc/0601121.

[45] D. Benincasa, In preparation.

[46] R. Sverdlov and L. Bombelli, *Class.Quant.Grav.* 26 075011 (2009).

[47] D. Dou and R. D. Sorkin, *Found. Phys.* 33 279 (2003).

[48] D. Dou, arXiv:gr-qc/0106024.

[49] D. Rideout and S. Zohren, *Class. Quant. Grav.* 23 6195 (2006).

[50] M. Ahmed, S. Dodelson, P. B. Greene and R. D. Sorkin, Everpresent lambda; *Phys. Rev. D* 69 103523 (2004).

[51] J. D. Barrow, Phys. Rev. D 75 067301 (2007), [arXiv:gr-qc/0612128], J. A. Zuntz, *Phys. Rev. D* 77 043002 (2008).

[52] F. Dowker, J. Henson and R. D. Sorkin, Mod. Phys. Lett. A19 1829-1840 (2004).

[53] R. D. Sorkin, *Ann. Phys.* 168 119-147 (1986).

[54] N. Kaloper and D. Mattingly, *Phys. Rev. D* 74 106001 (2006).

[55] D. Mattingly, *Phys. Rev. D* 77 125021 (2008).

[56] L. Philpott, F. Dowker and R. D. Sorkin, *Phys. Rev. D* 79 124047 (2009).

[57] L. Philpott, arXiv:1009.1593 [gr-qc].

[58] . R. Contaldi, F. Dowker and L. Philpott, *Class. Quant. Grav.* 27 172001 (2010).

[59] F. Dowker, J. Henson and R. Sorkin, arXiv:1009.3058 [gr-qc].

[60] R. Loll, *Living Rev. Rel.* 1 13 (1998).

[61] D. Kleitman and B. L. Rothschild, *Trans. Am. Math. Soc.* 205 205 (1975).

[62] D. Dhar, *J. Math. Phys.*, 19(8) 1711-1713 (1978).

[63] D. P. Rideout, arXiv:gr-qc/0212064.

[64] X. Martin, D. O'Connor, D. P. Rideout and R. D. Sorkin, *Phys. Rev. D* 63 084026 (2001).

[65] M. Ahmed and D. Rideout, *Phys. Rev. D* 81 083528 (2010).

[66] G. Brightwell, H. F. Dowker, R. S. Garcia, J. Henson and R. D. Sorkin, *Phys. Rev. D* 67 084031 (2003).

[67] F. Dowker, S. Johnston and S. Surya, arXiv:1007.2725 [gr-qc].

[68] R. D. Sorkin, Mod. Phys. Lett. A 9 3119 (1994).

[69] R. B. Salgado, *Mod. Phys. Lett. A* 17 711 (2002).

[70] S. Surya and P. Wallden, *Foundations of Physics* DOI: 10.1007/s10701-010-9419-1.

[71] F. Dowker, S. Johnston and R. D. Sorkin, *J. Phys. A* 43 275302 (2010).

[72] J. Henson, *Stud. Hist. Philos. Mod. Phys.* 36 519-543 (2005);

[73] J. Henson, arXiv:1102.2855[quant-ph]

[74] R. D. Sorkin, *J. Phys. A* 40 3207 (2007).

[75] F. Dowker and Y. Ghazi-Tabatabai, *J. Phys. A* 41 205306 (2008).

[76] R. D. Sorkin, arXiv:0911.1479 [gr-qc].

[77] J. Henson, D. Rideout, R. D. Sorkin and S. Surya, Work in progress.

[78] G. Brightwell, J. Henson and S. Surya, *Class. Quant. Grav.* 25 105025 (2008).

[79] M. H. El-Zahar and N. W. Sauer, *Order* 5 239 (1988).

[80] P. Winkler, *Order* 7 329 (1991).

[81] N. Acharyya and S. Vaidya, *JHEP* 1009 045 (2010).

[82] S. Johnston, arXiv:1010.5514 [hep-th].

In: Recent Research in Quantum Gravity ISBN: 978-1-61942-386-2
Editor: A. Dasgupta, pp. 41-94

Chapter 3

THE ASYMPTOTIC SAFETY PROGRAM IN QUANTUM EINSTEIN GRAVITY

Martin Reuter [1] ***and Frank Saueressig*** [2]
[1] Institute of Physics, University of Mainz, Staudingerweg, Mainz, Germany,
[2] Institut de Physique Théorique, CEA, IPhT, Gif-sur-Yvette, France CNRS, Gif-sur-Yvette, France,
and Institut des Hautes Études Scientifiques (IHES), Bures-sur-Yvette , France

Abstract

We review the renormalization group flow of Quantum Einstein Gravity (QEG) obtained within the effective average action approach. The possibility of constructing a fundamental theory of quantum gravity in the framework of asymptotic safety is discussed in detail, and the evidence supporting the viability of this program is summarized.

1. Introduction

After the introduction of a functional renormalization group equation for gravity [1] detailed investigations of the non-perturbative renormalization group (RG) behavior of Quantum Einstein Gravity have become possible [1]-[21]. The exact RG equation underlying this approach defines a Wilsonian RG flow on a theory space which consists of all diffeomorphism invariant functionals of the metric $g_{\mu\nu}$. The approach turned out to be ideal for investigating the asymptotic safety scenario in gravity [22, 23, 24] and, in fact, substantial evidence was found for the non-perturbative renormalizability of Quantum Einstein Gravity. The theory emerging from this construction (henceforth denoted "QEG") is not a quantization of classical General Relativity. Instead, its bare action corresponds to a non-trivial fixed point of the RG flow and is a prediction therefore. Independent support for the asymptotic safety conjecture comes from a two-dimensional symmetry reduction of the gravitational path-integral [25].

The approach of [1] employs the effective average action [26, 27, 28, 29] which has crucial advantages as compared to other continuum implementations of the Wilsonian RG flow [30]. In particular it is closely related to the standard effective action and defines a family of effective field theories $\{\Gamma_k[g_{\mu\nu}], 0 \leq k < \infty\}$ labeled by the coarse graining scale k. The latter property opens the door to a rather direct extraction of physical information from the RG flow, at least in single-scale cases: If the physical process under consideration involves a single typical momentum scale p_0 only, it can be described by a tree-level evaluation of $\Gamma_k[g_{\mu\nu}]$, with $k = p_0$.[1]

The effective field theory techniques proved useful for an understanding of the scale dependent geometry of the effective QEG spacetimes [31, 32, 33]. In particular it has been shown [3, 5, 33] that these spacetimes have fractal properties, with a fractal dimension of 2 at small, and 4 at large distances. The same dynamical dimensional reduction was also observed in numerical studies of Lorentzian dynamical triangulations [34, 35, 36]; in [37] A. Connes et al. speculated about its possible relevance to the non-commutative geometry of the standard model.

As for possible physics implications of the RG flow predicted by QEG, ideas from particle physics, in particular the "RG improvement", have been employed in order to study the leading quantum gravity effects in black hole and cosmological spacetimes [38]-[48]. Among other results, it was found [38] that the quantum effects tend to decrease the Hawking temperature of black holes, and that their evaporation process presumably stops completely once the black holes

[1] The precision which can be achieved by this effective field theory description depends on the size of the fluctuations relative to mean values. If they turn out large, or if more than one scale is involved, it might be necessary to go beyond the tree-level analysis.

mass is of the order of the Planck mass.

This review is intended to provide the background necessary for understanding these developments. In the next section we introduce the general idea of the effective average action and its associated functional renormalization group equation (FRGE) by means of a simple scalar example [26, 28], before reviewing the corresponding construction for gravity [1] in section 3. In all practical calculations based upon this approach which have been performed to date the truncation of theory space has been used as a non-perturbative approximation scheme. In section 3 we explain the general ideas and problems behind this method, and in section 4 we illustrate it explicitly in a simple context, the so-called Einstein-Hilbert truncation. Section 5 introduces the concept of asymptotic safety while section 6 contains a summary of the results obtained using truncated flow equations, with an emphasis on the question as to whether there exists a non-trivial fixed point for the average action's RG flow. If so, QEG could be established as a fundamental theory of quantum gravity which is non-perturbatively renormalizable and "asymptotically safe" from unphysical divergences.

2. Introducing the Effective Average Action

In this section we introduce the concept of the effective average action [26, 28, 27, 29] in the simplest context: scalar field theory on flat d-dimensional Euclidean space $\mathbb{R}^d$.

2.1. The Basic Construction for Scalar Fields

We start by considering a single-component real scalar field χ: $\mathbb{R}^d \rightarrow \mathbb{R}$ whose dynamics is governed by the bare action $S[\chi]$. Typically the functional S has the structure $S[\chi] = \int d^dx \left\{ \frac{1}{2}(\partial_\mu \chi)^2 + \frac{1}{2}m^2\chi^2 + \text{interactions} \right\}$, but we shall not need to assume any specific form of S in the following. After coupling $\chi(x)$ to a source $J(x)$ we can write down an a priori formal path integral representation for the generating functional of the connected Green's functions: $W[J] = \ln \int \mathcal{D}\chi \exp\{-S[\chi] + \int d^dx\, \chi(x)J(x)\}$. By definition, the (conventional) effective action $\Gamma[\phi]$ is the Legendre transform of $W[J]$. It depends on the field expectation value $\phi \equiv \langle \chi \rangle = \delta W[J]/\delta J$ and generates all 1-particle irreducible Greens functions of the theory by multiple functional differentiation with respect to $\phi(x)$ and setting $\phi = \phi[J = 0]$ thereafter. In order to make the functional integral well-defined a UV cutoff is needed; for example one could replace $\mathbb{R}^d$ by a d-dimensional lattice $\mathbb{Z}^d$. The functional integral $\mathcal{D}\chi$ would then read $\prod_{x \in \mathbb{Z}^d} d\chi(x)$. In the following we implicitly assume such a UV regularization but leave the details unspecified and use continuum notation for the fields and their Fourier transforms.

The construction of the effective average action [26] starts out from a modified form, $W_k[J]$, of the functional $W[J]$ which depends on a variable mass scale k. This scale is used to separate the Fourier modes of χ into "short wave length" and "long wave length", depending on whether or not their momentum square $p^2 \equiv p_\mu p^\mu$ is larger or smaller than k^2. By construction, the modes with $p^2 > k^2$ contribute without any suppression to the functional integral defining $W_k[J]$, while those with $p^2 < k^2$ contribute only with a reduced weight or are suppressed altogether, depending on which variant of the formalism is used. The new functional $W_k[J]$ is obtained from the conventional one by adding a "cutoff action" $\Delta_k S[\chi]$ to the bare action $S[\chi]$:

$$\exp\{W_k[J]\} = \int \mathcal{D}\chi \exp\Big\{ - S[\chi] - \Delta_k S[\chi] + \int d^dx\, \chi(x) J(x) \Big\} . \tag{1}$$

The factor $\exp\{-\Delta_k S[\chi]\}$ serves the purpose of suppressing the "IR modes" having $p^2 < k^2$. In momentum space the cutoff action is taken to be of the form

$$\Delta_k S[\chi] \equiv \frac{1}{2} \int \frac{d^dp}{(2\pi)^d} \mathcal{R}_k(p^2)\, |\widehat{\chi}(p)|^2 , \tag{2}$$

where $\widehat{\chi}(p) = \int d^dx\, \chi(x) \exp(-ipx)$ is the Fourier transform of $\chi(x)$. The precise shape of the function $\mathcal{R}_k(p^2)$ is arbitrary to some extent; what matters is its limiting behavior for $p^2 \gg k^2$ and $p^2 \ll k^2$ only. In the simplest case[2] we require that

$$\mathcal{R}_k(p^2) \approx \begin{cases} k^2 & \text{for } p^2 \ll k^2 , \\ 0 & \text{for } p^2 \gg k^2 . \end{cases} \tag{3}$$

The first condition leads to a suppression of the small momentum modes by a soft mass-like IR cutoff, the second guarantees that the large momentum modes

[2] We shall discuss a slight generalization of these conditions at the end of this section.

are integrated out in the usual way. Adding $\Delta_k S$ to the bare action $S[\chi]$ leads to

$$S + \Delta_k S = \frac{1}{2}\int \frac{d^d p}{(2\pi)^d}\Big[p^2 + m^2 + \mathcal{R}_k(p^2)\Big]\,|\widehat{\chi}(p)|^2 + \text{interactions}\,. \quad (4)$$

Obviously the cutoff function $\mathcal{R}_k(p^2)$ has the interpretation of a momentum dependent mass square which vanishes for $p^2 \gg k^2$ and assumes the constant value k^2 for $p^2 \ll k^2$. How $\mathcal{R}_k(p^2)$ is assumed to interpolate between these two regimes is a matter of calculational convenience. In practical calculations one often uses the exponential cutoff $\mathcal{R}_k(p^2) = p^2[\exp(p^2/k^2) - 1]^{-1}$, but many other choices are possible [28, 49]. One could also think of suppressing the $p^2 < k^2$ modes completely. This could be achieved by allowing $\mathcal{R}_k(p^2)$ to diverge for $p^2 \ll k^2$ so that $\exp\{-\Delta_k S[\chi]\} \to 0$ for modes with $p^2 \ll k^2$. While this behavior of $\mathcal{R}_k(p^2)$ seems most natural from the viewpoint of a Kadanoff-Wilson type coarse graining, its singular behavior makes the resulting generating functional problematic to deal with technically. For this reason, and since it still allows for the derivation of an exact RG equation, one usually prefers to work with a smooth cutoff satisfying (3). At the non-perturbative path integral level it suppresses the long wavelength modes by a factor $\exp\{-\frac{1}{2}k^2\int|\widehat{\chi}|^2\}$. In perturbation theory, according to eq. (4), the $\Delta_k S$ term leads to the modified propagator $[p^2 + m^2 + \mathcal{R}_k(p^2)]^{-1}$ which equals $[p^2 + m^2 + k^2]^{-1}$ for $p^2 \ll k^2$. Thus, when computing loops with this propagator, k^2 acts indeed as a conventional IR cutoff if $m^2 \ll k^2$. (It plays no role in the opposite limit $m^2 \gg k^2$ in which the physical particle mass cuts off the p-integration.) We note that by replacing p^2 with $-\partial^2$ in the argument of $\mathcal{R}_k(p^2)$ the cutoff action can be written in a way which makes no reference to the Fourier decomposition of χ:

$$\Delta_k S[\chi] = \frac{1}{2}\int d^d x\, \chi(x)\mathcal{R}_k(-\partial^2)\chi(x)\,. \quad (5)$$

The next steps towards the definition of the effective average action are similar to the usual procedure. One defines the (now k-dependent) field expectation value $\phi(x) \equiv \langle\chi(x)\rangle = \delta W_k[J]/\delta J(x)$, assumes that the functional relationship $\phi = \phi[J]$ can be inverted to yield $J = J[\phi]$, and introduces the Legendre transform of W_k,

$$\widetilde{\Gamma}_k[\phi] \equiv \int d^d x\, J(x)\phi(x) - W_k[J]\,, \quad (6)$$

where $J = J[\phi]$. The actual effective average action, denoted by $\Gamma_k[\phi]$, is obtained from $\widetilde{\Gamma}_k$ by subtracting $\Delta_k S[\phi]$:

$$\Gamma_k[\phi] \equiv \widetilde{\Gamma}_k[\phi] - \frac{1}{2}\int dx\, \phi(x)\mathcal{R}_k(-\partial^2)\phi(x)\,. \tag{7}$$

The rationale for this definition becomes clear when we look at the list of properties enjoyed by the functional Γ_k:

(1) The scale dependence of Γ_k is governed by the FRGE

$$k\frac{\partial}{\partial k}\Gamma_k[\phi] = \frac{1}{2}\mathrm{Tr}\Big[k\frac{\partial}{\partial k}\mathcal{R}_k\left(\Gamma_k^{(2)}[\phi] + \mathcal{R}_k\right)^{-1}\Big]\,. \tag{8}$$

Here the RHS uses a compact matrix notation. In a position space representation $\Gamma_k^{(2)}$ has the matrix elements $\Gamma_k^{(2)}(x,y) \equiv \delta^2\Gamma_k/\delta\phi(x)\delta\phi(y)$, i.e., it is the Hessian of the average action, $\mathcal{R}_k(x,y) \equiv \mathcal{R}_k(-\partial_x^2)\delta(x-y)$, and the trace Tr corresponds to an integral $\int d^dx$. In (8) the implicit UV cutoff can be removed trivially. This is most easily seen in the momentum representation where $k\frac{\partial}{\partial k}\mathcal{R}_k(p^2)$, considered a function of p^2, is significantly different from zero only in the region centered around $p^2 = k^2$. Hence the trace receives contributions from a thin shell of momenta $p^2 \approx k^2$ only and is therefore well convergent both in the UV and IR.

The RHS of (8) can be rewritten in a style reminiscent of a one-loop expression:

$$k\frac{\partial}{\partial k}\Gamma_k[\phi] = \frac{1}{2}\frac{D}{D\ln k}\mathrm{Tr}\ln\left(\Gamma_k^{(2)}[\phi] + \mathcal{R}_k\right)\,. \tag{9}$$

Here the scale derivative $D/D\ln k$ acts only on the k-dependence of $\mathcal{R}_k$, not on $\Gamma_k^{(2)}$. The $\mathrm{Tr}\ln(\cdots) = \ln\det(\cdots)$ expression in (9) differs from a standard one-loop determinant in two ways: it contains the Hessian of the actual effective action rather than that of the bare action S and it has a built in IR regulator $\mathcal{R}_k$. These modifications make (9) an *exact* equation. In a sense, solving it amounts to solving the complete theory.

The derivation of (8) proceeds as follows [26]. Taking the k-derivative of (6) with (1) and (5) inserted one finds

$$k\frac{\partial}{\partial k}\widetilde{\Gamma}_k[\phi] = \frac{1}{2}\int d^dx d^dy\, \langle\chi(x)\chi(y)\rangle\, k\frac{\partial}{\partial k}\mathcal{R}_k(x,y)\,, \tag{10}$$

with $\langle A\rangle \equiv e^{-W_k}\int\mathcal{D}\chi\, A\exp\{-S-\Delta_k S-\int J\phi\}$ defining the J and k dependent expectation values. Next it is convenient to introduce the connected 2-point function $G_{xy}\equiv G(x,y)\equiv \delta^2 W_k[J]/\delta J(x)\delta J(y)$ and the Hessian of $\widetilde{\Gamma}_k$: $(\widetilde{\Gamma}_k^{(2)})_{xy}\equiv \delta^2\widetilde{\Gamma}_k[J]/\delta\phi(x)\delta\phi(y)$. Since W_k and $\widetilde{\Gamma}_k$ are related by a Legendre transformation one shows in the usual way that G and $\widetilde{\Gamma}^{(2)}$ are mutually inverse matrices: $G\widetilde{\Gamma}^{(2)}=1$. Furthermore, taking two J-derivatives of (1) one obtains $\langle\chi(x)\chi(y)\rangle = G(x,y)+\phi(x)\phi(y)$. Substituting this expression for the two-point function into (10) we arrive at

$$\partial_t\widetilde{\Gamma}_k[\phi]=\frac{1}{2}\mathrm{Tr}[\partial_t\mathcal{R}_k G]+\frac{1}{2}\int d^dx\,\phi(x)\,\partial_t\mathcal{R}_k(-\partial^2)\,\phi(x)\,, \tag{11}$$

where $t\equiv\ln(k/k_0)$. In terms of Γ_k, the effective average action proper, this becomes $\partial_t\Gamma_k[\phi]=\frac{1}{2}\mathrm{Tr}[\partial_t\mathcal{R}_k G]$. The cancellation of the $\frac{1}{2}\int\phi\mathcal{R}_k\phi$ term is a first motivation for the definition (7) where this term is subtracted from the Legendre transform $\widetilde{\Gamma}_k$. The derivation is completed by noting that $G=[\widetilde{\Gamma}^{(2)}]^{-1}=(\Gamma_k^{(2)}+\mathcal{R}_k)^{-1}$, where the second equality follows by differentiating (7): $\Gamma_k^{(2)}=\widetilde{\Gamma}_k^{(2)}-\mathcal{R}_k$.

(2) The effective average action satisfies the following integro-differential equation:

$$\begin{aligned}\exp\{-\Gamma_k[\phi]\} &= \int\mathcal{D}\chi\,\exp\Big\{-S[\chi]+\int d^dx\,(\chi-\phi)\frac{\delta\Gamma_k[\phi]}{\delta\phi}\Big\}\\ &\quad\times\exp\Big\{-\int d^dx\,(\chi-\phi)\mathcal{R}_k(-\partial^2)(\chi-\phi)\Big\}.\end{aligned} \tag{12}$$

This equation is easily derived by combining eqs. (1), (6) and (7), and by using the effective field equation $\delta\widetilde{\Gamma}_k/\delta\phi = J$, which is 'dual' to $\delta W_k/\delta J=\phi$. (Note that it is $\widetilde{\Gamma}_k$ which appears here, not Γ_k.)

(3) For $k\to 0$ the effective average action approaches the ordinary effective action, $\lim_{k\to 0}\Gamma_k=\Gamma$, and for $k\to\infty$ the bare action $\Gamma_{k\to\infty}=S$. The $k\to 0$ limit is a consequence of (3), $\mathcal{R}_k(p^2)$ vanishes for all $p^2>0$ when $k\to 0$. The derivation of the $k\to\infty$ limit makes use of the integro-differential equation (12). A formal version the argument is as follows. Since $\mathcal{R}_k(p^2)$ approaches k^2 for $k\to\infty$, the second exponential on the RHS of (12) becomes $\exp\{-k^2\int dx(\chi-\phi)^2\}$, which, up to a normalization factor, approaches a delta-functional $\delta[\chi-\phi]$. The χ integration can be performed trivially then and one ends up with $\lim_{k\to\infty}\Gamma_k[\phi] = S[\phi]$. In a more careful treatment [26] one shows that the saddle point approximation of the functional integral in (12) about the point $\chi = \phi$ becomes exact in the limit $k \to \infty$. As a result, $\lim_{k\to\infty}\Gamma_k$ and S differ at most by the infinite mass limit of a one-loop determinant, which we suppress here since it plays no role in typical applications (see [50] for a more detailed discussion).

(4) The FRGE (8) is independent of the bare action S which enters only via the initial condition $\Gamma_\infty = S$. In the FRGE approach the calculation of the path integral for W_k is replaced by integrating the RG equation from $k = \infty$, where the initial condition $\Gamma_\infty = S$ is imposed, down to $k = 0$, where the effective average action equals the ordinary effective action Γ, the object which we actually would like to know.

2.2. Theory Space

The arena in which the Wilsonian RG dynamics takes place is the "theory space". Albeit a somewhat formal notion it helps in visualizing various concepts related to functional renormalization group equations, see fig. 1. To describe it, we shall be slightly more general than in the previous subsection and consider an arbitrary set of fields $\phi(x)$. Then the corresponding theory space consists of all (action) functionals $A : \phi \mapsto A[\phi]$ depending on this set, possibly subject to certain symmetry requirements (a $\mathbb{Z}_2$-symmetry for a single scalar, or diffeomorphism invariance if ϕ denotes the spacetime metric, for instance). So the theory space $\{A[\,\cdot\,]\}$ is fixed once the field content and the symmetries are fixed. Let us assume we can find a set of "basis functionals" $\{P_\alpha[\,\cdot\,]\}$ so that every point of theory space has an expansion of the form [23]

$$A[\phi] = \sum_{\alpha=1}^{\infty} \bar{u}_\alpha \, P_\alpha[\phi] \tag{13}$$

The basis $\{P_\alpha[\,\cdot\,]\}$ will include both local field monomials and non-local invariants and we may use the "generalized couplings" $\{\bar{u}_\alpha, \alpha = 1, 2, \dots\}$ as local coordinates. More precisely, the theory space is coordinatized by the subset of "essential couplings", i.e., those coordinates which cannot be absorbed by a field reparameterization.

Geometrically speaking the FRGE for the effective average action, eq. (8) or its generalization for an arbitrary set of fields, defines a vector field $\vec{\beta}$ on theory space. The integral curves along this vector field are the "RG trajectories" $k \mapsto \Gamma_k$ parameterized by the scale k. They start, for $k \to \infty$, at the bare action S (up to the correction term mentioned earlier) and terminate at the ordinary effective action at $k = 0$. The natural orientation of the trajectories is from higher to lower scales k, the direction of increasing "coarse graining". Expanding Γ_k as in (13),

$$\Gamma_k[\phi] = \sum_{\alpha=1}^{\infty} \overline{u}_\alpha(k)\, P_\alpha[\phi]\,, \tag{14}$$

the trajectory is described by infinitely many "running couplings" $\overline{u}_\alpha(k)$. Inserting (14) into the FRGE we obtain a system of infinitely many coupled differential equations for the $\overline{u}_\alpha$'s:

$$k\partial_k\, \overline{u}_\alpha(k) = \overline{\beta}_\alpha(\overline{u}_1, \overline{u}_2, \ldots; k)\,, \quad \alpha = 1, 2, \ldots. \tag{15}$$

Here the "beta functions" $\overline{\beta}_\alpha$ arise by expanding the trace on the RHS of the FRGE in terms of $\{P_\alpha[\,\cdot\,]\}$, i.e., $\frac{1}{2}\mathrm{Tr}\,[\cdots] = \sum_{\alpha=1}^{\infty} \overline{\beta}_\alpha(\overline{u}_1, \overline{u}_2, \cdots; k) P_\alpha[\phi]$. The expansion coefficients $\overline{\beta}_\alpha$ have the interpretation of beta functions similar to those of perturbation theory, but not restricted to relevant couplings. In standard field theory jargon one would refer to $\overline{u}_\alpha(k = \infty)$ as the "bare" parameters and to $\overline{u}_\alpha(k = 0)$ as the "renormalized" or "dressed" parameters.

The notation with the bar on $\overline{u}_\alpha$ and $\overline{\beta}_\alpha$ is to indicate that we are still dealing with dimensionful couplings. Usually the flow equation is reexpressed in terms of the dimensionless couplings $u_\alpha \equiv k^{-d_\alpha}\overline{u}_\alpha$, where d_α is the canonical mass dimension of $\overline{u}_\alpha$. Correspondingly the essential u_α's are used as coordinates of theory space. The resulting RG equations

$$k\partial_k u_\alpha(k) = \beta_\alpha(u_1, u_2, \ldots) \tag{16}$$

are a coupled system of autonomous differential equations. The β_α's have no explicit k-dependence and define a "time independent" vector field on theory space.

Fig. 1 gives a schematic summary of the theory space and its structures. It should be kept in mind, though, that only the essential couplings are coordinates on theory space, and that Γ_∞ and S might differ by a simple, explicitly known functional.

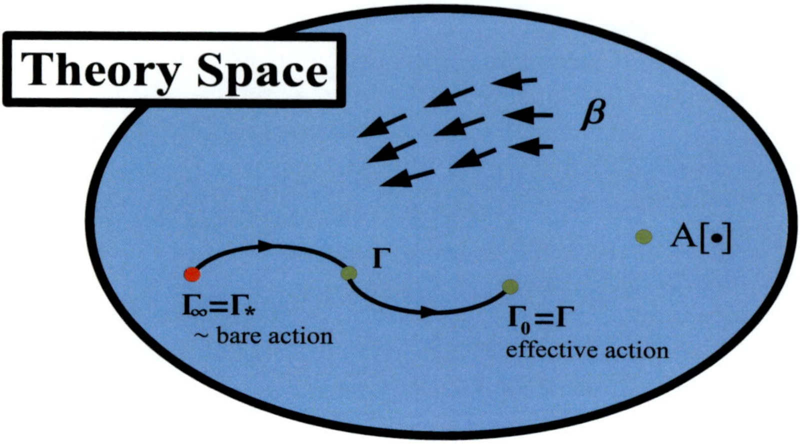

Figure 1. The points of theory space are the action functionals $A[\,\cdot\,]$. The RG equation defines a vector field $\vec{\beta}$ on this space; its integral curves are the RG trajectories $k \mapsto \Gamma_k$. They emanate from the fixed point action Γ_* and end at the standard effective action Γ.

2.3. Non-perturbative Approximations through Truncations

Up to this point our discussion did not involve any approximation. In practice, however, it is usually impossible to find exact solutions to the flow equation. As a way out, one could evaluate the trace on the RHS of the FRGE by expanding it with respect to some small coupling constant, for instance, thus recovering the familiar perturbative beta functions. A more interesting option which gives rise to non-perturbative approximate solutions is to truncate the theory space $\{A[\,\cdot\,]\}$. The basic idea is to project the RG flow onto a finite dimensional subspace of theory space. The subspace should be chosen in such a way that the projected flow encapsulates the essential physical features of the exact flow on the full space.

Concretely the projection onto a truncation subspace is performed as follows. One makes an ansatz of the form $\Gamma_k[\phi] = \sum_{i=1}^{N} \overline{u}_i(k) P_i[\phi]$, where the k-independent functionals $\{P_i[\,\cdot\,], i = 1, \ldots, N\}$ form a 'basis' on the subspace selected. For a scalar field, say, examples include pure potential terms

$\int d^d x \phi^m(x)$, $\int d^d x \phi^n(x) \ln \phi^2(x)$, ..., a standard kinetic term $\int d^d x (\partial\phi)^2$, higher order derivative terms $\int d^d x\, \phi\, (\partial^2)^n\, \phi$, $\int d^d x\, f(\phi)\, (\partial^2)^n\, \phi\, (\partial^2)^m\, \phi$, ..., and non-local terms like $\int d^d x\, \phi \ln(-\partial^2)\phi, \ldots$. Even if $S = \Gamma_\infty$ is simple, a standard ϕ^4 action, say, the evolution from $k = \infty$ downwards will generate such terms, a priori only constrained by symmetry requirements. The difficult task in practical RG applications consists in selecting a set of P_i's which, on the one hand, is generic enough to allow for a sufficiently precise description of the physics one is interested in, and which, on the other hand, is small enough to be computationally manageable.

The projected RG flow is described by a set of ordinary (if $N < \infty$) differential equations for the couplings $\overline{u}_i(k)$. They arise as follows. Let us assume we expand the ϕ-dependence of $\frac{1}{2}\mathrm{Tr}[\cdots]$ (with the ansatz for $\Gamma_k[\phi]$ inserted) in a basis $\{P_\alpha[\,\cdot\,]\}$ of the *full* theory space which contains the P_i's spanning the truncated space as a subset:

$$\frac{1}{2}\mathrm{Tr}[\cdots] = \sum_{\alpha=1}^{\infty} \overline{\beta}_\alpha(\overline{u}_1, \ldots, \overline{u}_N; k)\, P_\alpha[\phi]$$
$$= \sum_{i=1}^{N} \overline{\beta}_i(\overline{u}_1, \ldots, \overline{u}_N; k)\, P_i[\phi] + \text{rest}\,. \tag{17}$$

Here the "rest" contains all terms outside the truncated theory space; the approximation consists in neglecting precisely those terms. Thus, equating (17) to the LHS of the flow equation, $\partial_t \Gamma_k = \sum_{i=1}^{N} \partial_t \overline{u}_i(k) P_i$, the linear independence of the P_i's implies the coupled system of ordinary differential equations

$$\partial_t \overline{u}_i(k) = \overline{\beta}_i(\overline{u}_1, \ldots, \overline{u}_N; k)\,, \quad i = 1, \ldots, N\,. \tag{18}$$

Solving (18) one obtains an *approximation* to the exact RG trajectory projected onto the chosen subspace. Note that this approximate trajectory does, in general, not coincide with the projection of the exact trajectory, but if the subspace is well chosen, it will not be very different from it. In fact, the most non-trivial problem in using truncated flow equations is to find and justify a truncation subspace which should be as low dimensional as possible to make the calculations feasible, but at the same time large enough to describe at least qualitatively the essential physics. We shall return to the issue of testing the quality of a given truncation later on.

As a simple example of a truncation we mention the ‘local potential approximation’ [28]. The corresponding subspace consists of functionals containing a standard kinetic term plus arbitrary non-derivative terms:

$$\Gamma_k[\phi] \equiv \int d^dx \left\{ \frac{1}{2}(\partial\phi(x))^2 + U_k(\phi(x)) \right\}. \tag{19}$$

In this case N is infinite, the coordinates $\overline{u}_i$ on truncated theory space being the infinitely many parameters characterizing an arbitrary potential function $\phi \mapsto U(\phi)$. The infinitely many component equations (18) amount to a partial differential equation for the running potential $U_k(\phi)$. It is obtained by inserting (19) into the FRGE and projecting the trace onto functionals of the form (19). This is most easily done by inserting a *constant* field $\phi = \varphi = const$ into both sides of the equation since this gives a non-vanishing value precisely to the non-derivative P_i's. Since $\Gamma_k^{(2)} = -\partial^2 + U_k''(\varphi)$ with $U'' \equiv d^2U_k/d\phi^2$ has no explicit x-dependence the trace is easily evaluated in momentum space. This leads to the following partial differential equation:

$$k\partial_k U_k(\varphi) = \frac{1}{2}\int \frac{d^dp}{(2\pi)^d} \frac{k\partial_k \mathcal{R}_k(p^2)}{p^2 + \mathcal{R}_k(p^2) + U_k''(\varphi)}. \tag{20}$$

It describes how the classical (or microscopic) potential $U_\infty = V_{\text{class}}$ evolves into the standard effective potential $U_0 = V_{\text{eff}}$. Remarkably, the limit $\lim_{k\to 0} U_k$ is automatically a convex function of φ, and there is no need to perform the Maxwell construction ‘by hand’, in the case of spontaneous symmetry breaking. For a detailed discussion of this point we refer to [28].

One can continue the truncation process and make a specific ansatz for the φ-dependence of the running potential, $U_k(\varphi) = \frac{1}{2}\overline{m}(k)^2\varphi^2 + \frac{1}{12}\overline{\lambda}(k)\varphi^4$, say. Then, upon inserting $U_k''(\varphi) = \overline{m}(k)^2 + \overline{\lambda}(k)\varphi^2$ into the RHS of (19) and expanding to $O(\varphi^4)$ one can equate the coefficients of φ^2 and φ^4 to obtain the flow equations on a 2-dimensional subspace: $k\partial_k \overline{m}^2 = \overline{\beta}_{\overline{m}^2}$, $k\partial_k\overline{\lambda} = \overline{\beta}_{\overline{\lambda}}$.

If one wants to go beyond the local potential approximation (19) the first step is to allow for a (ϕ independent in the simplest case) wave function renormalization, i.e., a running prefactor of the kinetic term: $\Gamma_k = \int d^dx \left\{\frac{1}{2}Z_k(\partial\phi)^2 + U_k\right\}$. Using truncations of this type one should employ a slightly different normalization of $\mathcal{R}_k(p^2)$, namely $\mathcal{R}_k(p^2) \approx Z_k k^2$ for $p^2 \ll k^2$. Then $\mathcal{R}_k$ combines with $\Gamma_k^{(2)}$ to the inverse propagator $\Gamma_k^{(2)} + \mathcal{R}_k =$

$Z_k(p^2 + k^2) + \cdots$, as it is necessary if the IR cutoff is to give rise to a $(\mathrm{mass})^2$ of size k^2 rather than k^2/Z_k. In particular in more complicated theories with more than one field it is important that all fields are cut off at precisely the same k^2. This is achieved by a cutoff function of the form

$$\mathcal{R}_k(p^2) = \mathcal{Z}_k\, k^2\, R^{(0)}(p^2/k^2)\,, \tag{21}$$

where $R^{(0)}$ is normalized such that $R^{(0)}(0) = 1$ and $R^{(0)}(\infty) = 0$. In general the factor $\mathcal{Z}_k$ is a matrix in field space. In the sector of modes with inverse propagator $Z_k^{(i)} p^2 + \cdots$ this matrix is chosen diagonal with entries $\mathcal{Z}_k = Z_k^{(i)}$.

3. The Effective Average Action for Gravity

We saw that the FRGE of the effective average action does not depend on the bare action S. Given a theory space, the form of the FRGE and, as a result, the vector field $\vec{\beta}$ are completely fixed. To define a theory space $\{A[\,\cdot\,]\}$ one has to specify on which types of fields the functionals A are supposed to depend, and what their symmetries are. This is the only input data needed for finding the RG flow.

In the case of QEG the theory space consists, by definition, of functionals $A[g_{\mu\nu}]$ depending on a symmetric tensor field, the metric, in a diffeomorphism invariant way. Unfortunately it is not possible to straightforwardly apply the constructions of the previous section to this theory space. Diffeomorphism invariance leads to two types of complications one has to deal with [1].

The first one is not specific to the RG approach. It occurs already in the standard functional integral quantization of gauge or gravity theories, and is familiar from Yang-Mills theories. If one gauge-fixes the functional integral with an ordinary (covariant) gauge fixing condition like $\partial^\mu A^a_\mu = 0$, couples the (non-abelian) gauge field A^a_μ to a source, and constructs the ordinary effective action the resulting functional $\Gamma[A^a_\mu]$ is *not* invariant under the gauge transformations of A^a_μ, $A^a_\mu \mapsto A^a_\mu + D^{ab}_\mu(A)\,\omega^b$. Only at the level of physical quantities constructed from $\Gamma[A^a_\mu]$, S-matrix elements for instance, gauge invariance is recovered.

The second problem is related to the fact that in a gauge theory a "coarse graining" based on a naive Fourier decomposition of $A^a_\mu(x)$ is not gauge covariant and hence not physical. In fact, if one were to gauge transform a slowly varying $A^a_\mu(x)$ using a parameter function $\omega^a(x)$ with a fast x-variation, a gauge field with a fast x-variation would arise which, however, still describes the same physics. In a non-gauge theory the coarse graining is performed by expanding the field in terms of eigenfunctions of the (positive) operator $-\partial^2$ and declaring its eigenmodes 'long' or 'short' wavelength depending on whether the corresponding eigenvalue p^2 is smaller or larger than a given k^2. In a gauge theory the best one can do in installing this procedure is to expand with respect to the *covariant* Laplacian or a similar operator, and then organize the modes according to the size of their eigenvalues. While gauge covariant, this approach sacrifices to some extent the intuition of a Fourier coarse graining in terms of slow and fast modes. Analogous remarks apply to theories of gravity covariant under general coordinate transformations.

The key idea which led to a solution of both problems was the use of the background field method. In fact, it is well kown [51, 52] that the background gauge fixing method leads to an effective action which depends on its arguments in a gauge invariant way. As it turned out [27, 1] this technique also lends itself for implementing a covariant IR cutoff, and it is at the core of the effective average action for Yang-Mills theories [27, 29] and for gravity [1]. In the following we briefly review the effective average action for gravity which has been introduced in ref. [1].

The ultimate goal is to give meaning to an integral over 'all' metrics $\gamma_{\mu\nu}$ of the form $\int \mathcal{D}\gamma_{\mu\nu} \exp\{-S[\gamma_{\mu\nu}] + \text{source terms}\}$ whose bare action $S[\gamma_{\mu\nu}]$ is invariant under general coordinate transformations,

$$\delta\gamma_{\mu\nu} = \mathcal{L}_v \gamma_{\mu\nu} \equiv v^\rho \partial_\rho \gamma_{\mu\nu} + \partial_\mu v^\rho \gamma_{\rho\nu} + \partial_\nu v^\rho \gamma_{\rho\mu} \,, \tag{1}$$

where $\mathcal{L}_v$ is the Lie derivative with respect to the vector field $v^\mu \partial_\mu$. To start with we consider $\gamma_{\mu\nu}$ to be a Riemannian metric and assume that $S[\gamma_{\mu\nu}]$ is positive definite. Heading towards the background field formalism, the first step consists in decomposing the variable of integration according to $\gamma_{\mu\nu} = \bar{g}_{\mu\nu} + h_{\mu\nu}$, where $\bar{g}_{\mu\nu}$ is a fixed background metric. Note that we are not implying a perturbative expansion here, $h_{\mu\nu}$ is not supposed to be small in any sense. After the background split the measure $\mathcal{D}\gamma_{\mu\nu}$ becomes $\mathcal{D}h_{\mu\nu}$ and the gauge transformations which we have to gauge-fix read

$$\delta h_{\mu\nu} = \mathcal{L}_v \gamma_{\mu\nu} = \mathcal{L}_v(\bar{g}_{\mu\nu} + h_{\mu\nu}) \,, \quad \delta \bar{g}_{\mu\nu} = 0 \,. \tag{2}$$

Picking an a priori arbitrary gauge fixing condition $F_\mu(h;\overline{g}) = 0$ the Faddeev-Popov trick can be applied straightforwardly [51]. Upon including an IR cutoff as in the scalar case we are lead to the following k-dependent generating functional W_k for the connected Green functions:

$$\begin{aligned} &\exp\left\{W_k[t^{\mu\nu},\sigma^\mu,\bar{\sigma}_\mu;\overline{g}_{\mu\nu}]\right\} \\ &= \int \mathcal{D}h_{\mu\nu}\mathcal{D}C^\mu\mathcal{D}\bar{C}_\mu \exp\Big\{-S[\bar{g}+h] - S_{\rm gf}[h;\bar{g}] \\ &- S_{\rm gh}[h,C,\bar{C};\bar{g}] - \Delta_k S[h,C,\bar{C};\bar{g}] - S_{\rm source}\Big\}. \end{aligned} \tag{3}$$

Here $S_{\rm gf}$ denotes the gauge fixing term

$$S_{\rm gf}[h;\overline{g}] = \frac{1}{2\alpha}\int d^dx\sqrt{\overline{g}}\,\overline{g}^{\mu\nu}F_\mu F_\nu\,, \tag{4}$$

and $S_{\rm gh}$ is the action for the corresponding Faddeev–Popov ghosts C^μ and $\bar{C}_\mu$:

$$S_{\rm gh}[h,C,\bar{C};\overline{g}] = -\kappa^{-1}\int d^dx\,\bar{C}_\mu\,\overline{g}^{\mu\nu}\,\frac{\partial F_\nu}{\partial h_{\alpha\beta}}\,\mathcal{L}_C\left(\overline{g}_{\alpha\beta}+h_{\alpha\beta}\right). \tag{5}$$

The Faddeev–Popov action $S_{\rm gh}$ is obtained along the same lines as in Yang–Mills theory: one applies a gauge transformation (2) to F_μ and replaces the parameters v^μ by the ghost field C^μ. The integral over C^μ and $\bar{C}_\mu$ exponentiates the Faddeev-Popov determinant $\det[\delta F_\mu/\delta v^\nu]$. In (3) we coupled $h_{\mu\nu}$, C^μ and $\bar{C}_\mu$ to sources $t^{\mu\nu}$, $\bar{\sigma}_\mu$ and σ^μ, respectively: $S_{\rm source} = -\int d^dx\,\sqrt{\overline{g}}\Big\{t^{\mu\nu}h_{\mu\nu} + \bar{\sigma}_\mu C^\mu + \sigma^\mu\bar{C}_\mu\Big\}$. The k and source dependent expectation values of $h_{\mu\nu}$, C^μ and $\bar{C}_\mu$ are then given by

$$\bar{h}_{\mu\nu} = \frac{1}{\sqrt{\overline{g}}}\frac{\delta W_k}{\delta t^{\mu\nu}}, \qquad \xi^\mu = \frac{1}{\sqrt{\overline{g}}}\frac{\delta W_k}{\delta\bar{\sigma}_\mu}, \qquad \bar{\xi}_\mu = \frac{1}{\sqrt{\overline{g}}}\frac{\delta W_k}{\delta\sigma^\mu}. \tag{6}$$

As usual we assume that one can invert the relations (6) and solve for the sources $(t^{\mu\nu}\,,\,\sigma^\mu\,,\,\bar{\sigma}_\mu)$ as functionals of $(\bar{h}_{\mu\nu}\,,\,\xi^\mu\,,\,\bar{\xi}_\mu)$ and, parametrically, of $\overline{g}_{\mu\nu}$. The Legendre transform $\widetilde{\Gamma}_k$ of W_k reads

$$\widetilde{\Gamma}_k[\bar{h},\xi,\bar{\xi};\overline{g}] = \int d^dx\,\sqrt{\overline{g}}\left\{t^{\mu\nu}\bar{h}_{\mu\nu}+\bar{\sigma}_\mu\xi^\mu+\sigma^\mu\bar{\xi}_\mu\right\} - W_k[t,\sigma,\bar{\sigma};\overline{g}]\,. \tag{7}$$

This functional inherits a parametric $\overline{g}_{\mu\nu}$-dependence from W_k.

As mentioned earlier for a generic gauge fixing condition the Legendre transform (7) is not a diffeomorphism invariant functional of its arguments since the gauge breaking under the functional integral is communicated to $\widetilde{\Gamma}_k$ via the sources. While $\widetilde{\Gamma}_k$ does indeed describe the correct 'on-shell' physics satisfying all constraints coming from BRST invariance, it is not invariant off-shell [51, 52]. The situation is different for the class of gauge fixing conditions of the background type. While – as any gauge fixing condition must – they break the invariance under (2) they are chosen to be invariant under the so-called background gauge transformations

$$\delta h_{\mu\nu} = \mathcal{L}_v h_{\mu\nu} \,, \quad \delta \overline{g}_{\mu\nu} = \mathcal{L}_v \overline{g}_{\mu\nu} \,. \tag{8}$$

The complete metric $\gamma_{\mu\nu} = \overline{g}_{\mu\nu} + h_{\mu\nu}$ transforms as $\delta\gamma_{\mu\nu} = \mathcal{L}_v \gamma_{\mu\nu}$ both under (8) and under (2). The crucial difference is that the ('quantum') gauge transformations (2) keep $\overline{g}_{\mu\nu}$ unchanged so that the entire change of $\gamma_{\mu\nu}$ is ascribed to $h_{\mu\nu}$. This is the point of view one adopts in a standard perturbative calculation around flat space where one fixes $\overline{g}_{\mu\nu} = \eta_{\mu\nu}$ and allows for no variation of the background. In the present construction, instead, we leave $\overline{g}_{\mu\nu}$ unspecified but insist on covariance under (8). This will lead to a completely background covariant formulation.

Clearly there exist many possible gauge fixing terms $S_{\rm gf}[h;\overline{g}]$ of the form (4) which break (2) and are invariant under (8). A convenient choice which has been employed in practical calculations is the one-parameter family of gauge conditions

$$F_\mu[h;\overline{g}] = \sqrt{2}\,\kappa\,\big(\overline{D}^\nu\, h_{\mu\nu} - \frac{1+\varpi}{d}\,\overline{D}_\mu\, h^\nu{}_\nu\big)\,, \tag{9}$$

parameterized by ϖ. The covariant derivative $\bar{D}_\mu$ involves the Christoffel symbols $\bar{\Gamma}^\rho_{\mu\nu}$ of the background metric. Note that (9) is linear in the quantum field $h_{\alpha\beta}$. For $\varpi = \frac{1}{2}(d-2)$, (9) reduces to the background version of the harmonic coordinate condition [51]: on a flat background with $\overline{g}_{\mu\nu} = \eta_{\mu\nu}$ the condition $F_\mu = 0$ becomes the familiar harmonic coordinate condition, $\partial^\mu h_{\mu\nu} = \frac{1}{2}\partial_\nu h_\mu^{\;\mu}$. In eqs. (9) and (5) κ is an arbitrary constant with the dimension of a mass. We shall set $\kappa \equiv (32\pi\bar{G})^{-1/2}$ with $\bar{G}$ a constant reference value of Newton's constant. The ghost action for the gauge condition (9) reads

$$S_{\rm gh}[h, C, \bar{C}; \overline{g}] = -\sqrt{2}\int d^dx\,\sqrt{\overline{g}}\,\bar{C}_\mu \mathcal{M}[g,\overline{g}]^\mu{}_\nu C^\nu \tag{10}$$

with the Faddeev–Popov operator

$$\mathcal{M}[g,\overline{g}]^{\mu}{}_{\nu} = \overline{D}^{\rho}\, g^{\mu}{}_{\nu}\, D_{\rho} + \overline{D}^{\rho}\, g_{\rho\nu} D^{\mu} - 2\,\frac{1+\varpi}{d}\, \overline{D}_{\mu}\, \overline{g}^{\rho\sigma}\, g_{\rho\nu} D_{\sigma}\,. \qquad (11)$$

It will prove crucial that for every background-type choice of F_{μ}, S_{gh} is invariant under (8) together with

$$\delta C^{\mu} = \mathcal{L}_{v} C^{\mu}\,, \quad \delta \bar{C}_{\mu} = \mathcal{L}_{v} \bar{C}_{\mu}\,. \qquad (12)$$

The essential piece in eq. (3) is the IR cutoff for the gravitational field $h_{\mu\nu}$ and for the ghosts. It is taken to be of the form

$$\begin{aligned}\Delta_k S = &\frac{\kappa^2}{2} \int d^d x\, \sqrt{\bar{g}}\, h_{\mu\nu} \mathcal{R}_k^{\mathrm{grav}}[\bar{g}]^{\mu\nu\rho\sigma}\, h_{\rho\sigma} \\ &+ \sqrt{2} \int d^d x\, \sqrt{\bar{g}}\, \bar{C}_{\mu} \mathcal{R}_k^{\mathrm{gh}}[\bar{g}] C^{\mu}\,.\end{aligned} \qquad (13)$$

The cutoff operators $\mathcal{R}_k^{\mathrm{grav}}$ and $\mathcal{R}_k^{\mathrm{gh}}$ serve the purpose of discriminating between high–momentum and low–momentum modes. Eigenmodes of $-\bar{D}^2$ with eigenvalues $p^2 \gg k^2$ are integrated out without any suppression whereas modes with small eigenvalues $p^2 \ll k^2$ are suppressed. The operators $\mathcal{R}_k^{\mathrm{grav}}$ and $\mathcal{R}_k^{\mathrm{gh}}$ have the structure $\mathcal{R}_k[\bar{g}] = \mathcal{Z}_k k^2 R^{(0)}(-\bar{D}^2/k^2)$, where the dimensionless function $R^{(0)}$ interpolates between $R^{(0)}(0) = 1$ and $R^{(0)}(\infty) = 0$. A convenient choice is, e.g., the exponential cutoff $R^{(0)}(w) = w[\exp(w) - 1]^{-1}$ where $w = p^2/k^2$. The factors $\mathcal{Z}_k$ are different for the graviton and the ghost cutoff. For the ghost $\mathcal{Z}_k \equiv Z_k^{\mathrm{gh}}$ is a pure number, whereas for the metric fluctuation $\mathcal{Z}_k \equiv \mathcal{Z}_k^{\mathrm{grav}}$ is a tensor, constructed only from the background metric $\overline{g}_{\mu\nu}$, which must be fixed along the lines described at the end of section 2

A feature of $\Delta_k S$ which is essential from a practical point of view is that the modes of $h_{\mu\nu}$ and the ghosts are organized according to their eigenvalues with respect to the *background* Laplace operator $\bar{D}^2 = \overline{g}^{\mu\nu} \bar{D}_{\mu} \bar{D}_{\nu}$ rather than $D^2 = g^{\mu\nu} D_{\mu} D_{\nu}$, which would pertain to the full quantum metric $\overline{g}_{\mu\nu} + h_{\mu\nu}$. Using $\bar{D}^2$ the functional $\Delta_k S$ is quadratic in the quantum field $h_{\mu\nu}$, while it becomes extremely complicated if D^2 is used instead. The virtue of a quadratic $\Delta_k S$ is that it gives rise to a flow equation which contains only *second* functional derivatives of Γ_k but no higher ones. The flow equations resulting from the cutoff operator D^2 are prohibitively complicated and can hardly be used for practical computations. A second property of $\Delta_k S$ which is crucial for our purposes is that it is invariant under the background gauge transformations (8) with (12).

Having specified all the ingredients which enter the functional integral (3) for the generating functional W_k we can write down the final definition of the effective average action Γ_k. It is obtained from the Legendre transform $\widetilde{\Gamma}_k$ by subtracting the cutoff action $\Delta_k S$ with the classical fields inserted:

$$\Gamma_k[\bar{h}, \xi, \bar{\xi}; \bar{g}] = \widetilde{\Gamma}_k[\bar{h}, \xi, \bar{\xi}; \bar{g}] - \Delta_k S[\bar{h}, \xi, \bar{\xi}; \bar{g}] \, . \tag{14}$$

It is convenient to define the expectation value of the quantum metric $\gamma_{\mu\nu}$,

$$g_{\mu\nu}(x) \equiv \bar{g}_{\mu\nu}(x) + \bar{h}_{\mu\nu}(x) \, , \tag{15}$$

and consider Γ_k as a functional of $g_{\mu\nu}$ rather than $\bar{h}_{\mu\nu}$:

$$\Gamma_k[g_{\mu\nu}, \bar{g}_{\mu\nu}, \xi^\mu, \bar{\xi}_\mu] \equiv \Gamma_k[g_{\mu\nu} - \bar{g}_{\mu\nu}, \xi^\mu, \bar{\xi}_\mu; \bar{g}_{\mu\nu}] \, . \tag{16}$$

So, what did we gain going through this seemingly complicated background field construction, eventually ending up with an action functional which depends on *two* metrics even? The main advantage of this setting is that the corresponding functionals $\widetilde{\Gamma}_k$, and as a result Γ_k, are invariant under general coordinate transformations where all its arguments transform as tensors of the corresponding rank:

$$\Gamma_k[\Phi + \mathcal{L}_v \Phi] = \Gamma_k[\Phi] \, , \qquad \Phi \equiv \left\{ g_{\mu\nu}, \bar{g}_{\mu\nu}, \xi^\mu, \bar{\xi}_\mu \right\} \, . \tag{17}$$

Note that in (17), contrary to the "quantum gauge transformation" (2), also the background metric transforms as an ordinary tensor field: $\delta \bar{g}_{\mu\nu} = \mathcal{L}_v \bar{g}_{\mu\nu}$. Eq. (17) is a consequence of

$$W_k \left[\mathcal{J} + \mathcal{L}_v \mathcal{J} \right] = W_k \left[\mathcal{J} \right] \, , \qquad \mathcal{J} \equiv \left\{ t^{\mu\nu}, \sigma^\mu, \bar{\sigma}_\mu; \bar{g}_{\mu\nu} \right\} \, . \tag{18}$$

This invariance property follows from (3) if one performs a compensating transformation (8), (12) on the integration variables $h_{\mu\nu}$, C^μ and $\bar{C}_\mu$ and uses the invariance of $S[\bar{g} + h]$, S_{gf}, S_{gh} and $\Delta_k S$. At this point we assume that the functional measure in (3) is diffeomorphism invariant.

Since the $\mathcal{R}_k$'s vanish for $k = 0$, the limit $k \to 0$ of $\Gamma_k[g_{\mu\nu}, \overline{g}_{\mu\nu}, \xi^\mu, \bar{\xi}_\mu]$ brings us back to the standard effective action functional which still depends on two metrics, though. The "ordinary" effective action $\Gamma[g_{\mu\nu}]$ with one metric argument is obtained from this functional by setting $\overline{g}_{\mu\nu} = g_{\mu\nu}$, or equivalently $\bar{h}_{\mu\nu} = 0$ [51, 52]:

$$\begin{aligned}\Gamma[g] &\equiv \lim_{k\to 0} \Gamma_k[g, \overline{g} = g, \xi = 0, \bar{\xi} = 0] \\ &= \lim_{k\to 0} \Gamma_k[\bar{h} = 0, \xi = 0, \bar{\xi} = 0; g = \overline{g}] \,.\end{aligned} \tag{19}$$

This equation brings about the "magic property" of the background field formalism: a priori the 1PI n-point functions of the metric are obtained by an n-fold functional differentiation of $\Gamma_0[\bar{h}, 0, 0; \overline{g}_{\mu\nu}]$ with respect to $\bar{h}_{\mu\nu}$. Hereby $\overline{g}_{\mu\nu}$ is kept fixed; it acts simply as an externally prescribed function which specifies the form of the gauge fixing condition. Hence the functional Γ_0 and the resulting *off-shell* Green functions do depend on $\overline{g}_{\mu\nu}$, but the *on-shell* Green functions, related to observable scattering amplitudes, do not depend on $\overline{g}_{\mu\nu}$. In this respect $\overline{g}_{\mu\nu}$ plays a role similar to the gauge parameter α in the standard approach. Remarkably, the same on-shell Green functions can be obtained by differentiating the functional $\Gamma[g_{\mu\nu}]$ of (19) with respect to $g_{\mu\nu}$, or equivalently $\Gamma_0[\bar{h} = 0, \xi = 0, \bar{\xi} = 0; \overline{g} = g]$, with respect to its $\overline{g}$ argument. In this context, 'on-shell' means that the metric satisfies the effective field equation $\delta\Gamma_0[g]/\delta g_{\mu\nu} = 0$.

With (19) and its k-dependent counterpart

$$\bar{\Gamma}_k[g_{\mu\nu}] \equiv \Gamma_k[g_{\mu\nu}, g_{\mu\nu}, 0, 0] \tag{20}$$

we succeeded in constructing a diffeomorphism invariant generating functional for gravity: thanks to (17) $\Gamma[g_{\mu\nu}]$ and $\bar{\Gamma}_k[g_{\mu\nu}]$ are invariant under general coordinate transformations $\delta g_{\mu\nu} = \mathcal{L}_v g_{\mu\nu}$. However, there is a price to be paid for their invariance: the simplified functional $\bar{\Gamma}_k[g_{\mu\nu}]$ does not satisfy an exact RG equation, basically because it contains insufficient information. The actual RG evolution has to be performed at the level of the functional $\Gamma_k[g, \overline{g}, \xi, \bar{\xi}]$. Only *after* the evolution one may set $\overline{g} = g$, $\xi = 0, \bar{\xi} = 0$. As a result, the actual theory space of QEG, $\{A[g, \overline{g}, \xi, \bar{\xi}]\}$, consists of functionals of all four variables, $g_{\mu\nu}, \overline{g}_{\mu\nu}, \xi^\mu, \bar{\xi}_\mu$, subject to the invariance condition (17).

The derivation of the FRGE for Γ_k is analogous to the scalar case. Following exactly the same steps one arrives at

$$\partial_t \Gamma_k[\bar{h}, \xi, \bar{\xi}; \bar{g}] = \frac{1}{2} \mathrm{Tr}\left[\left(\Gamma_k^{(2)} + \widehat{\mathcal{R}}_k\right)^{-1}_{\bar{h}\bar{h}} \left(\partial_t \widehat{\mathcal{R}}_k\right)_{\bar{h}\bar{h}}\right] - \frac{1}{2} \mathrm{Tr}\left[\left\{\left(\Gamma_k^{(2)} + \widehat{\mathcal{R}}_k\right)^{-1}_{\bar{\xi}\xi} - \left(\Gamma_k^{(2)} + \widehat{\mathcal{R}}_k\right)^{-1}_{\xi\bar{\xi}}\right\} \left(\partial_t \widehat{\mathcal{R}}_k\right)_{\bar{\xi}\xi}\right] . \tag{21}$$

Here $\Gamma_k^{(2)}$ denotes the Hessian of Γ_k with respect to the dynamical fields $\bar{h}$, ξ, $\bar{\xi}$ at fixed $\bar{g}$. It is a block matrix labeled by the fields $\varphi_i \equiv \{\bar{h}_{\mu\nu}, \xi^\mu, \bar{\xi}_\mu\}$:

$$\Gamma_k^{(2)\,ij}(x,y) \equiv \frac{1}{\sqrt{\bar{g}(x)\bar{g}(y)}} \frac{\delta^2 \Gamma_k}{\delta\varphi_i(x)\delta\varphi_j(y)} . \tag{22}$$

(In the ghost sector the derivatives are understood as left derivatives.) Likewise, $\widehat{\mathcal{R}}_k$ is a block diagonal matrix with entries $(\widehat{\mathcal{R}}_k)_{\bar{h}\bar{h}}^{\mu\nu\rho\sigma} \equiv \kappa^2 (\mathcal{R}_k^{\mathrm{grav}}[\bar{g}])^{\mu\nu\rho\sigma}$ and $\widehat{\mathcal{R}}_{\bar{\xi}\xi} = \sqrt{2}\mathcal{R}_k^{\mathrm{gh}}[\bar{g}]$. Performing the trace in the position representation it includes an integration $\int d^d x \sqrt{\bar{g}(x)}$ involving the background volume element. For any cutoff which is qualitatively similar to the exponential cutoff the traces on the RHS of eq. (21) are well convergent, both in the IR and the UV. By virtue of the factor $\partial_t \widehat{\mathcal{R}}_k$, the dominant contributions come from a narrow band of generalized momenta centered around k. Large momenta are exponentially suppressed.

Besides the FRGE the effective average action also satisfies an exact integro-differential equation similar to (12) in the scalar case. By the same argument as there it can be used to find the $k \to \infty$ limit of the average action:

$$\Gamma_{k\to\infty}[\bar{h}, \xi, \bar{\xi}; \bar{g}] = S[\bar{g} + \bar{h}] + S_{\mathrm{gf}}[\bar{h}; \bar{g}] + S_{\mathrm{gh}}[\bar{h}, \xi, \bar{\xi}; \bar{g}] . \tag{23}$$

Note that the 'initial value' $\Gamma_{k\to\infty}$ includes the gauge fixing and ghost actions. At the level of the functional $\bar{\Gamma}_k[g]$, eq. (23) boils down to $\bar{\Gamma}_{k\to\infty}[g] = S[g]$. However, as $\Gamma_k^{(2)}$ involves derivatives with respect to $\bar{h}_{\mu\nu}$ (or equivalently $g_{\mu\nu}$) at fixed $\bar{g}_{\mu\nu}$ it is clear that the evolution cannot be formulated entirely in terms of $\bar{\Gamma}_k$ alone.

The background gauge invariance of Γ_k, expressed in eq. (17), is of enormous practical importance. It implies that if the initial functional does not contain non-invariant terms, the flow will not generate such terms. Very often this reduces the number of terms to be retained in a reliable truncation ansatz quite considerably. Nevertheless, even if the initial action is simple, the RG flow will generate all sorts of local and non-local terms in Γ_k which are consistent with the symmetries.

Let us close this section by remarking that, at least formally, the construction of the effective average action can be repeated for Lorentzian signature metrics. In this case one deals with oscillating exponentials e^{iS}, and for arguments like the one leading to (23) one has to employ the Riemann-Lebesgue lemma. Apart from the obvious substitutions $\Gamma_k \to -i\Gamma_k$, $\mathcal{R}_k \to -i\mathcal{R}_k$ the evolution equation remains unaltered.

4. Truncated Flow Equations

Solving the FRGE (21) subject to the initial condition (23) is equivalent to (and in practice as difficult as) calculating the original functional integral over $\gamma_{\mu\nu}$. It is therefore important to devise efficient approximation methods. The truncation of theory space is the one which makes maximum use of the FRGE reformulation of the quantum field theory problem at hand.

As for the flow on the theory space $\{A[g, \overline{g}, \xi, \bar{\xi}]\}$ a still very general truncation consists of neglecting the evolution of the ghost action by making the ansatz

$$\Gamma_k[g, \overline{g}, \xi, \bar{\xi}] = \bar{\Gamma}_k[g] + \widehat{\Gamma}_k[g, \overline{g}] + S_{\rm gf}[g - \overline{g}; \overline{g}] + S_{\rm gh}[g - \overline{g}, \xi, \bar{\xi}; \overline{g}] \,, \quad (1)$$

where we extracted the classical $S_{\rm gf}$ and $S_{\rm gh}$ from Γ_k. The remaining functional depends on both $g_{\mu\nu}$ and $\overline{g}_{\mu\nu}$. It is further decomposed as $\bar{\Gamma}_k + \widehat{\Gamma}_k$ where $\bar{\Gamma}_k$ is defined as in (20) and $\widehat{\Gamma}_k$ contains the deviations for $\overline{g} \neq g$. Hence, by definition, $\widehat{\Gamma}_k[g, g] = 0$, and $\widehat{\Gamma}_k$ contains in particular quantum corrections to the gauge fixing term which vanishes for $\overline{g} = g$, too. This ansatz satisfies the initial condition (23) if

$$\bar{\Gamma}_{k\to\infty} = S \qquad \text{and} \qquad \widehat{\Gamma}_{k\to\infty} = 0 \,. \quad (2)$$

Inserting (1) into the exact FRGE (21) one obtains an evolution equation on the truncated space $\{A[g, \overline{g}]\}$:

$$\partial_t \Gamma_k[g, \overline{g}] = \frac{1}{2} \mathrm{Tr} \left[\left(\kappa^{-2} \Gamma_k^{(2)}[g, \overline{g}] + \mathcal{R}_k^{\rm grav}[\overline{g}] \right)^{-1} \partial_t \mathcal{R}_k^{\rm grav}[\overline{g}] \right]$$

$$- \operatorname{Tr}\left[\left(-\mathcal{M}[g,\bar{g}] + \mathcal{R}_k^{\text{gh}}[\bar{g}]\right)^{-1} \partial_t \mathcal{R}_k^{\text{gh}}[\bar{g}]\right] . \tag{3}$$

This equation evolves the functional

$$\Gamma_k[g,\bar{g}] \equiv \bar{\Gamma}_k[g] + S_{\text{gf}}[g-\bar{g};\bar{g}] + \widehat{\Gamma}_k[g,\bar{g}] . \tag{4}$$

Here $\Gamma_k^{(2)}$ denotes the Hessian of $\Gamma_k[g,\bar{g}]$ with respect to $g_{\mu\nu}$ at fixed $\bar{g}_{\mu\nu}$.

The truncation ansatz (1) is still too general for practical calculations to be easily possible. The first truncation for which the RG flow has been found [1] is the "Einstein-Hilbert truncation" which retains in $\bar{\Gamma}_k[g]$ only the terms $\int d^dx \sqrt{g}$ and $\int d^dx \sqrt{g}R$, already present in the classical action, with k-dependent coupling constants, and includes only the wave function renormalization in $\widehat{\Gamma}_k$:

$$\begin{aligned} \Gamma_k[g,\bar{g}] =& 2\kappa^2 Z_{Nk} \int d^dx \sqrt{g} \left\{-R(g) + 2\bar{\lambda}_k\right\} \\ &+ \frac{Z_{Nk}}{2\alpha} \int d^dx \sqrt{\bar{g}}\, \bar{g}^{\mu\nu} F_\mu F_\nu . \end{aligned} \tag{5}$$

In this case the truncation subspace is 2-dimensional. The ansatz (5) contains two free functions of the scale, the running cosmological constant $\bar{\lambda}_k$ and Z_{Nk} or, equivalently, the running Newton constant $G_k \equiv \bar{G}/Z_{Nk}$. Here $\bar{G}$ is a fixed constant, and $\kappa \equiv (32\pi\bar{G})^{-1/2}$. As for the gauge fixing term, F_μ is given by eq. (9) with $\bar{h}_{\mu\nu} \equiv g_{\mu\nu} - \bar{g}_{\mu\nu}$ replacing $h_{\mu\nu}$; it vanishes for $g = \bar{g}$. The ansatz (5) has the general structure of (1) with

$$\widehat{\Gamma}_k = (Z_{Nk} - 1) S_{\text{gf}} . \tag{6}$$

Within the Einstein-Hilbert approximation the gauge fixing parameter α is kept constant. Here we shall set $\alpha = 1$ and comment on generalizations later on.

Upon inserting the ansatz (5) into the flow equation (3) it boils down to a system of two ordinary differential equations for Z_{Nk} and $\bar{\lambda}_k$. Their derivation is rather technical, so we shall focus on the conceptual aspects here. In order to find $\partial_t Z_{Nk}$ and $\partial_t \bar{\lambda}_k$ it is sufficient to consider (3) for $g_{\mu\nu} = \bar{g}_{\mu\nu}$. In this case the LHS of the flow equation becomes $2\kappa^2 \int d^dx \sqrt{g}[-R(g)\partial_t Z_{Nk} + 2\partial_t(Z_{Nk}\bar{\lambda}_k)]$. The RHS is assumed to admit an expansion in terms of invariants $P_i[g_{\mu\nu}]$. In the Einstein-Hilbert truncation only two of them, $\int d^dx \sqrt{g}$ and $\int d^dx \sqrt{g}R$, need to be retained. They can be extracted from the traces in (3) by standard derivative expansion techniques. Equating the result to the LHS and comparing the coefficients of $\int d^dx\sqrt{g}$ and $\int d^dx\sqrt{g}R$, a pair of coupled differential equations for Z_{Nk} and λ_k arises. It is important to note that, on the RHS, we may set $g_{\mu\nu} = \overline{g}_{\mu\nu}$ only *after* the functional derivatives of $\Gamma_k^{(2)}$ have been obtained since they must be taken at fixed $\overline{g}_{\mu\nu}$.

In principle this calculation can be performed without ever considering any specific metric $g_{\mu\nu} = \overline{g}_{\mu\nu}$. This reflects the fact that the approach is background covariant. The RG flow is universal in the sense that it does not depend on any specific metric. In this respect gravity is not different from the more traditional applications of the renormalization group: the RG flow in the Ising universality class, say, has nothing to do with any specific spin configuration, it rather reflects the statistical properties of very many such configurations.

While there is no conceptual necessity to fix the background metric, it nevertheless is sometimes advantageous from a computational point of view to pick a specific class of backgrounds. Leaving $\overline{g}_{\mu\nu}$ completely general, the calculation of the functional traces is very hard work usually. In principle there exist well known derivative expansion and heat kernel techniques which could be used for this purpose, but their application is an extremely lengthy and tedious task usually. Moreover, typically the operators $\Gamma_k^{(2)}$ and $\mathcal{R}_k$ are of a complicated non-standard type so that no efficient use of the tabulated Seeley coefficients can be made. However, often calculations of this type simplify if one can assume that $g_{\mu\nu} = \overline{g}_{\mu\nu}$ has specific properties. Since the beta functions are background independent we may therefore restrict $\overline{g}_{\mu\nu}$ to lie in a conveniently chosen class of geometries which is still general enough to disentangle the invariants retained and at the same time simplifies the calculation.

For the Einstein-Hilbert truncation the most efficient choice is a family of d-spheres $S^d(r)$, labeled by their radius r. For those geometries, $D_\alpha R_{\mu\nu\rho\sigma} = 0$, so they give a vanishing value to all invariants constructed from $g = \overline{g}$ containing covariant derivatives acting on curvature tensors. What remains (among the local invariants) are terms of the form $\int\sqrt{g}P(R)$, where P is a polynomial in the Riemann tensor with arbitrary index contractions. To linear order in the (contractions of the) Riemann tensor the two invariants relevant for the Einstein-Hilbert truncation are discriminated by the S^d metrics as the latter scale differently with the radius of the sphere: $\int\sqrt{g} \sim r^d$, $\int\sqrt{g}R \sim r^{d-2}$. Thus, in order to compute the beta functions of $\bar{\lambda}_k$ and Z_{Nk} it is sufficient to insert an S^d metric with arbitrary r and to compare the coefficients of r^d and r^{d-2}. If one wants to do better and include the three quadratic invariants $\int R_{\mu\nu\rho\sigma}R^{\mu\nu\rho\sigma}$, $\int R_{\mu\nu}R^{\mu\nu}$, and $\int R^2$, the family $S^d(r)$ is not general enough to separate them; all scale like r^{d-4} with the radius.

Under the trace we need the operator $\Gamma_k^{(2)}[\bar{h};\bar{g}]$. It is most easily calculated by Taylor expanding the truncation ansatz, $\Gamma_k[\bar{g}+\bar{h},\bar{g}] = \Gamma_k[\bar{g},\bar{g}] + O(\bar{h}) + \Gamma_k^{\text{quad}}[\bar{h};\bar{g}] + O(\bar{h}^3)$, and stripping off the two $\bar{h}$'s from the quadratic term, $\Gamma_k^{\text{quad}} = \frac{1}{2}\int \bar{h}\Gamma_k^{(2)}\bar{h}$. For $\bar{g}_{\mu\nu}$ the metric on $S^d(r)$ one obtains

$$\begin{aligned}\Gamma_k^{\text{quad}}[\bar{h};\bar{g}] =& \frac{1}{2} Z_{Nk}\kappa^2 \int d^dx \sqrt{\bar{g}} \Big\{ \widehat{h}_{\mu\nu}\left[-\bar{D}^2 - 2\bar{\lambda}_k + C_T\bar{R}\right]\widehat{h}^{\mu\nu} \\ & - \left(\frac{d-2}{2d}\right)\phi\left[-\bar{D}^2 - 2\bar{\lambda}_k + C_S\bar{R}\right]\phi \Big\}, \end{aligned} \tag{7}$$

with $C_T \equiv (d(d-3)+4)/(d(d-1))$, $C_S \equiv (d-4)/d$. In order to partially diagonalize this quadratic form $\bar{h}_{\mu\nu}$ has been decomposed into a traceless part $\widehat{h}_{\mu\nu}$ and the trace part proportional to ϕ: $\bar{h}_{\mu\nu} = \widehat{h}_{\mu\nu} + d^{-1}\bar{g}_{\mu\nu}\phi$, $\bar{g}^{\mu\nu}\widehat{h}_{\mu\nu} = 0$. Further, $\bar{D}^2 = \bar{g}^{\mu\nu}\bar{D}_\mu\bar{D}_\nu$ is the covariant Laplace operator corresponding to the background geometry, and $\bar{R} = d(d-1)/r^2$ is the numerical value of the curvature scalar on $S^d(r)$.

At this point we can fix the constants $\mathcal{Z}_k$ which appear in the cutoff operators $\mathcal{R}_k^{\text{grav}}$ and $\mathcal{R}_k^{\text{gh}}$ of (13). They should be adjusted in such a way that for every low–momentum mode the cutoff combines with the kinetic term of this mode to $-\bar{D}^2 + k^2$ times a constant. Looking at (7) we see that the respective kinetic terms for $\widehat{h}_{\mu\nu}$ and ϕ differ by a factor of $-(d-2)/2d$. This suggests the following choice:

$$\left(\mathcal{Z}_k^{\text{grav}}\right)^{\mu\nu\rho\sigma} = \left[(\mathbb{I} - P_\phi)^{\mu\nu\rho\sigma} - \frac{d-2}{2d}P_\phi^{\mu\nu\rho\sigma}\right] Z_{Nk}\,. \tag{8}$$

Here $(P_\phi)_{\mu\nu}{}^{\rho\sigma} = d^{-1}\bar{g}_{\mu\nu}\bar{g}^{\rho\sigma}$ is the projector on the trace part of the metric. For the traceless tensor (8) gives $\mathcal{Z}_k^{\text{grav}} = Z_{Nk}\mathbb{I}$, and for ϕ the different relative normalization is taken into account. (See ref. [1] for a detailed discussion of the subtleties related to this choice.) Thus we obtain in the $\widehat{h}$ and the ϕ-sector, respectively:

$$
\begin{aligned}
&\left(\kappa^{-2}\Gamma_k^{(2)}[g,g]+\mathcal{R}_k^{\text{grav}}\right)_{\widehat{h}\widehat{h}} \\
&\qquad = Z_{Nk}\left[-D^2+k^2R^{(0)}(-D^2/k^2)-2\bar{\lambda}_k+C_T R\right], \\
&\left(\kappa^{-2}\Gamma_k^{(2)}[g,g]+\mathcal{R}_k^{\text{grav}}\right)_{\phi\phi} \\
&\qquad = -\frac{d-2}{2d}Z_{Nk}\left[-D^2+k^2R^{(0)}(-D^2/k^2)-2\bar{\lambda}_k+C_S R\right].
\end{aligned}
\tag{9}
$$

From now on we may set $\overline{g} = g$ and for simplicity we have omitted the bars from the metric and the curvature. Since we did not take into account any renormalization effects in the ghost action we set $Z_k^{\text{gh}} \equiv 1$ in $\mathcal{R}_k^{\text{gh}}$ and obtain

$$
-\mathcal{M}+\mathcal{R}_k^{\text{gh}} = -D^2+k^2R^{(0)}(-D^2/k^2)+C_V R\,, \tag{10}
$$

with $C_V \equiv -1/d$. At this point the operator under the first trace on the RHS of (3) has become block diagonal, with the $\widehat{h}\widehat{h}$ and $\phi\phi$ blocks given by (9). Both block operators are expressible in terms of the Laplacian D^2, in the former case acting on traceless symmetric tensor fields, in the latter on scalars. The second trace in (3) stems from the ghosts; it contains (10) with D^2 acting on vector fields.

It is now a matter of straightforward algebra to compute the first two terms in the derivative expansion of those traces, proportional to $\int d^dx\sqrt{g} \sim r^d$ and $\int d^dx\sqrt{g}R \sim r^{d-2}$. Considering the trace of an arbitrary function of the Laplacian, $W(-D^2)$, the expansion up to second order derivatives of the metric is given by

$$
\begin{aligned}
\text{Tr}[W(-D^2)] =& (4\pi)^{-d/2}\text{tr}(I)\Big\{Q_{d/2}[W]\int d^dx\,\sqrt{g} \\
&+\frac{1}{6}Q_{d/2-1}[W]\int d^dx\,\sqrt{g}R+O(R^2)\Big\}.
\end{aligned}
\tag{11}
$$

The Q_n's are defined as

$$
Q_n[W] = \frac{1}{\Gamma(n)}\int_0^\infty dz\, z^{n-1}W(z)\,, \tag{12}
$$

for $n > 0$, and $Q_0[W] = W(0)$ for $n = 0$. The trace $\mathrm{tr}(I)$ counts the number of independent field components. It equals 1, d, and $(d-1)(d+2)/2$, for scalars, vectors, and symmetric traceless tensors, respectively. The expansion (11) is easily derived using standard heat kernel and Mellin transform techniques [1].

Using (11) it is easy to calculate the traces in (3) and to obtain the RG equations in the form $\partial_t Z_{Nk} = \cdots$ and $\partial_t(Z_{Nk}\bar{\lambda}_k) = \cdots$. We shall not display them here since it is more convenient to rewrite them in terms of the dimensionless running cosmological constant and Newton constant, respectively:

$$\lambda_k \equiv k^{-2}\bar{\lambda}_k \,, \qquad g_k \equiv k^{d-2} G_k \equiv k^{d-2} Z_{Nk}^{-1} \bar{G} \,. \tag{13}$$

Recall that the dimensionful running Newton constant is given by $G_k = Z_{Nk}^{-1}\bar{G}$. In terms of the dimensionless couplings g and λ the RG equations become a system of autonomous differential equations:

$$\partial_t g_k = \left[d - 2 + \eta_N(g_k, \lambda_k)\right] g_k \equiv \beta_g(g_k, \lambda_k) \,, \tag{14a}$$

$$\partial_t \lambda_k = \beta_\lambda(g_k, \lambda_k) \,. \tag{14b}$$

Here $\eta_N \equiv -\partial_t \ln Z_{Nk}$ is the anomalous dimension of the operator $\sqrt{g}R$,

$$\eta_N(g_k, \lambda_k) = \frac{g_k \, B_1(\lambda_k)}{1 - g_k \, B_2(\lambda_k)} \,, \tag{15}$$

with the following functions of λ_k:

$$\begin{aligned} B_1(\lambda_k) \equiv & \frac{1}{3}(4\pi)^{1-d/2}\Big[d(d+1)\Phi^1_{d/2-1}(-2\lambda_k) - 6d(d-1)\Phi^2_{d/2}(-2\lambda_k) \\ & \qquad - 4d\Phi^1_{d/2-1}(0) - 24\Phi^2_{d/2}(0)\Big] \\ B_2(\lambda_k) \equiv & -\frac{1}{6}(4\pi)^{1-d/2}\Big[d(d+1)\widetilde{\Phi}^1_{d/2-1}(-2\lambda_k) - 6d(d-1)\widetilde{\Phi}^2_{d/2}(-2\lambda_k)\Big] \,. \end{aligned} \tag{16}$$

The beta function for λ is given by a similar expression:

$$\begin{aligned} \beta_\lambda(g_k, \lambda_k) = & -(2 - \eta_N)\lambda_k + \frac{1}{2} g_k (4\pi)^{1-d/2} \\ & \cdot \Big[2d(d+1)\Phi^1_{d/2}(-2\lambda_k) - 8d\Phi^1_{d/2}(0) - d(d+1)\eta_N \widetilde{\Phi}^1_{d/2}(-2\lambda_k)\Big] \,. \end{aligned} \tag{17}$$

The "threshold functions" Φ and $\widetilde{\Phi}$ appearing in (16) and (17) are certain integrals involving the normalized cutoff function $R^{(0)}$:

$$\begin{aligned}\Phi^p_n(w) &\equiv \frac{1}{\Gamma(n)}\int_0^\infty dz\, z^{n-1}\frac{R^{(0)}(z)-zR^{(0)\prime}(z)}{[z+R^{(0)}(z)+w]^p}\,,\\ \widetilde{\Phi}^p_n(w) &\equiv \frac{1}{\Gamma(n)}\int_0^\infty dz\, z^{n-1}\frac{R^{(0)}(z)}{[z+R^{(0)}(z)+w]^p}\,.\end{aligned} \tag{18}$$

They are defined for positive integers p, and $n > 0$.

With the derivation of the system (14) we managed to find an approximation to a two-dimensional projection of the RG flow. Its properties, and in particular the domain of applicability and reliability of the Einstein-Hilbert truncation will be discussed in the following section.

While there are (few) aspects of the truncated RG flow which are independent of the cutoff scheme, i.e., independent of the function $R^{(0)}$, the explicit solution of the flow equation requires a specific choice of this function. As we discussed already, the normalized cutoff function $R^{(0)}(w)$, $w = p^2/k^2$, describes the "shape" of $\mathcal{R}_k(p^2)$ in the transition region where it interpolates between the prescribed behavior for $p^2 \ll k^2$ and $p^2 \gg k^2$, respectively, and is referred to as the "shape function" therefore. In the literature various forms of $R^{(0)}$'s have been employed. Easy to handle, but disadvantageous for high precision calculations is the sharp cutoff [4] defined by $\mathcal{R}_k(p^2) = \lim_{\hat{R}\to\infty} \hat{R}\,\theta(1 - p^2/k^2)$, where the limit is to be taken after the p^2 integration. This cutoff allows for an evaluation of the Φ and $\widetilde{\Phi}$ integrals in closed form. Taking $d = 4$ as an example, eqs. (14) boil down to the following simple system of equations:[3]

$$\partial_t \lambda_k = -(2-\eta_N)\lambda_k - \frac{g_k}{\pi}\left[5\ln(1-2\lambda_k) - 2\zeta(3) + \frac{5}{2}\eta_N\right], \tag{19a}$$

$$\partial_t g_k = (2+\eta_N)\, g_k\,, \tag{19b}$$

$$\eta_N = -\frac{2\,g_k}{6\pi + 5\,g_k}\left[\frac{18}{1-2\lambda_k} + 5\ln(1-2\lambda_k) - \zeta(2) + 6\right]. \tag{19c}$$

Also the "optimized cutoff" [49] with $R^{(0)}(w) = (1-w)\theta(1-w)$ allows for an analytic evaluation of the integrals [14]. In order to check the scheme

[3]To be precise, (19) corresponds to the sharp cutoff with $s = 1$, see [4].

(in)dependence of the results it is desirable to perform the calculation for a whole class of $R^{(0)}$'s. For this purpose the following one parameter family of exponential cutoffs has been used [8, 3, 5]:

$$R^{(0)}(w;s) = \frac{sw}{e^{sw}-1}\,. \tag{20}$$

The precise form of the cutoff is controlled by the "shape parameter" s. For $s = 1$, (20) coincides with the standard exponential cutoff. The exponential cutoffs are suitable for precision calculations, but the price to be paid is that their Φ and $\widetilde{\Phi}$ integrals can be evaluated only numerically. The same is true for a one-parameter family of shape functions with compact support which was used in [3, 5].

Above we illustrated the general ideas and constructions underlying gravitational RG flows by means of the simplest example, the Einstein-Hilbert truncation. In the literature various extensions have been investigated. The derivation and analysis of these more general flow equations, corresponding to higher dimensional truncation subspaces, is an extremely complex and calculationally demanding problem in general. For this reason we cannot go into the technical details here and just mention some further developments.

(1) The natural next step beyond the Einstein-Hilbert truncation consists in generalizing the functional $\bar{\Gamma}_k[g]$, while keeping the gauge fixing and ghost sector classical, as in (1). During the RG evolution the flow generates all possible diffeomorphism invariant terms in $\bar{\Gamma}_k[g]$ which one can construct from $g_{\mu\nu}$. Both local and non-local terms are induced. The local invariants contain strings of curvature tensors and covariant derivatives acting upon them, with any number of tensors and derivatives, and of all possible index structures. The first truncation of this class which has been worked out completely [5, 6] is the "R^2-truncation" defined by (1) with the same $\widehat{\Gamma}_k$ as before, and the $(\text{curvature})^2$ action

$$\bar{\Gamma}_k[g] = \int d^dx\sqrt{g}\Big\{(16\pi G_k)^{-1}[-R(g)+2\bar{\lambda}_k] + \bar{\beta}_k R^2(g)\Big\}\,. \tag{21}$$

In this case the truncated theory space is 3-dimensional. Its natural (dimensionless) coordinates are (g, λ, β), where $\beta_k \equiv k^{4-d}\bar{\beta}_k$, and g and λ defined in (13). Even though (21) contains only one additional invariant, the derivation of the corresponding RG equations is far more complicated than in the Einstein-Hilbert case. We shall summarize the results obtained with (21) [5, 6] in the next section.

(2) As for generalizing the ghost sector of the truncation beyond (1) no results are available yet, but there is a partial result concerning the gauge fixing term. Even if one makes the ansatz (5) for $\Gamma_k[g, \bar{g}]$ in which the gauge fixing term has the classical (or more appropriately, bare) structure one should treat its prefactor as a running coupling: $\alpha = \alpha_k$. The beta function of α has not been determined yet from the FRGE, but there is a simple argument which allows us to bypass this calculation.

In non-perturbative Yang-Mills theory and in perturbative quantum gravity $\alpha = \alpha_k = 0$ is known to be a fixed point for the α evolution. The following reasoning suggests that the same is true within the non-perturbative FRGE approach to gravity. In the standard functional integral the limit $\alpha \to 0$ corresponds to a sharp implementation of the gauge fixing condition, i.e., $\exp(-S_{\text{gf}})$ becomes proportional to $\delta[F_\mu]$. The domain of the $\int \mathcal{D}h_{\mu\nu}$ integration consists of those $h_{\mu\nu}$'s which satisfy the gauge fixing condition exactly, $F_\mu = 0$. Adding the IR cutoff at k amounts to suppressing some of the $h_{\mu\nu}$ modes while retaining the others. But since all of them satisfy $F_\mu = 0$, a variation of k cannot change the domain of the $h_{\mu\nu}$ integration. The delta functional $\delta[F_\mu]$ continues to be present for any value of k if it was there originally. As a consequence, α vanishes for all k, i.e., $\alpha = 0$ is a fixed point of the α evolution [53].

Thus we can mimic the dynamical treatment of a running α by setting the gauge fixing parameter to the constant value $\alpha = 0$. The calculation for $\alpha = 0$ is more complicated than at $\alpha = 1$, but for the Einstein-Hilbert truncation the α-dependence of β_g and β_λ, for arbitrary constant α has been found in [54, 3]. The R^2-truncations could be analyzed only in the simple $\alpha = 1$ gauge, but the results from the Einstein-Hilbert truncation suggest the UV quantities of interest do not change much between $\alpha = 0$ and $\alpha = 1$ [3, 5].

(3) In refs. [19] and [20] gravitational RG flows have been explored in an approximation that goes beyond a truncation of theory space. Here only the subsector of the basic path integral over the conformal degrees of freedom has been considered while all others were omitted. This leads to a scalar-like theory to which the same apparatus underlying the analysis of the full theory has been applied (average action, background decomposition). Remarkably, this "conformally reduced gravity" (contrary to a standard 4-dimensional scalar theory) possesses a NGFP and a RG flow that is qualitatively similar to that of full QEG. It was possible to establish the existence of the NGFP on an *infinite dimensional* theory space consisting of arbitrary potentials for the conformal factor. These somewhat unexpected results find their explanation [19, 20] by noting that the quantization scheme based upon the (conformal reduction of the) gravitational average action is *"background independent"* in the sense that no special metric (flat space, etc.) plays a distinguished role.

(4) Up to now we considered pure gravity. As for as the general formalism, the inclusion of matter fields is straightforward. The structure of the flow equation remains unaltered, except that now $\Gamma_k^{(2)}$ and $\mathcal{R}_k$ are operators on the larger Hilbert space of both gravity and matter fluctuations. In practice the derivation of the projected RG equations can be quite a formidable task, however, the difficult part being the decoupling of the various modes (diagonalization of $\Gamma_k^{(2)}$) which in most calculational schemes is necessary for the computation of the functional traces. Various matter systems, both interacting and non-interacting (apart from their interaction with gravity) have been studied in the literature [2, 55, 56]. A rather detailed analysis has been performed by Percacci et al. In [2, 12] arbitrary multiplets of free (massless) fields with spin $0, 1/2, 1$ and $3/2$ were included. In [12] an interacting scalar theory coupled to gravity in the Einstein-Hilbert approximation was analyzed, and a possible solution to the triviality and the hierarchy problem [57] was proposed in this context.

(5) Finally we mention another generalization of the simplest case reviewed above which is of a more technical nature [3]. In order to facilitate the calculation of the functional traces it is helpful to employ a transverse-traceless (TT) decomposition of the metric:

$$h_{\mu\nu} = h^T_{\mu\nu} + \bar{D}_\mu V_\nu + \bar{D}_\nu V_\mu + \bar{D}_\mu \bar{D}_\nu \sigma - d^{-1} \overline{g}_{\mu\nu} \bar{D}^2 \sigma + d^{-1} \overline{g}_{\mu\nu} \phi. \quad (22)$$

Here $h^T_{\mu\nu}$ is a transverse traceless tensor, V_μ a transverse vector, and σ and ϕ are scalars. In this framework it is natural to formulate the cutoff in terms of the component fields appearing in the TT decomposition: $\Delta_k S \sim \int h^T_{\mu\nu} \mathcal{R}_k h^{T\mu\nu} + \int V_\mu \mathcal{R}_k V^\mu + \cdots$. This cutoff is referred to as a cutoff of "type B", in contradistinction to the "type A" cutoff described above, $\Delta_k S \sim \int h_{\mu\nu} \mathcal{R}_k h^{\mu\nu}$. Since covariant derivatives do not commute the two cutoffs are not exactly equal even if they contain the same shape function. Thus, comparing type A and type B cutoffs is an additional possibility for checking scheme (in)dependence [3, 5].

5. Asymptotic Safety

In intuitive terms, the basic idea of asymptotic safety can be understood as follows. The boundary of theory space depicted in fig. 1 is meant to separate points with coordinates $\{u_\alpha, \alpha = 1, 2, \ldots\}$ with all the essential couplings u_α well defined, from points with undefined, divergent couplings. The basic task of renormalization theory consists in constructing an "infinitely long" RG trajectory which lies entirely within this theory space, i.e., a trajectory which neither leaves theory space (that is, develops divergences) in the UV limit $k \to \infty$ nor in the IR limit $k \to 0$. Every such trajectory defines one possible quantum theory.

The idea of asymptotic safety is to perform the UV limit $k \to \infty$ at a fixed point $\{u^*_\alpha, \alpha = 1, 2, \ldots\} \equiv u^*$ of the RG flow. The fixed point is a zero of the vector field $\vec{\beta} \equiv (\beta_\alpha)$, i.e., $\beta_\alpha(u^*) = 0$ for all $\alpha = 1, 2, \ldots$. The RG trajectories, solutions of $k\partial_k u_\alpha(k) = \beta_\alpha(u(k))$, have a low "velocity" near a fixed point because the β_α's are small there and directly at the fixed point the running stops completely. As a result, one can "use up" an infinite amount of RG time near/at the fixed point if one bases the quantum theory on a trajectory which runs into such a fixed point for $k \to \infty$. This is the key idea of asymptotic safety: If in the UV limit the trajectory ends at a fixed point, an "inner point" of theory space giving rise to a well behaved action functional, we can be sure that, for $k \to \infty$, the trajectory does not escape from theory space, i.e., does not develop pathological properties such as divergent couplings. For $k \to \infty$ the resulting quantum theory is "asymptotically safe" from unphysical divergences. In the context of gravity, Weinberg [22] proposed to use a non-Gaussian fixed point (NGFP) for letting $k \to \infty$. By definition, not all of its coordinates u^*_α vanish.[4]

Recall from section 2.2 that the coordinates u_α are the *dimensionless* essential couplings related to the dimensionful ones $\bar{u}_\alpha$ by $u_\alpha \equiv k^{-d_\alpha}\bar{u}_\alpha$. Hence the

[4]In contrast, $u^*_\alpha = 0, \forall\alpha = 1, 2, \ldots$ is a so-called Gaussian fixed point (GFP). In a sense standard perturbation theory takes the $k \to \infty$ limit at the GFP; see [23] for a detailed discussion.

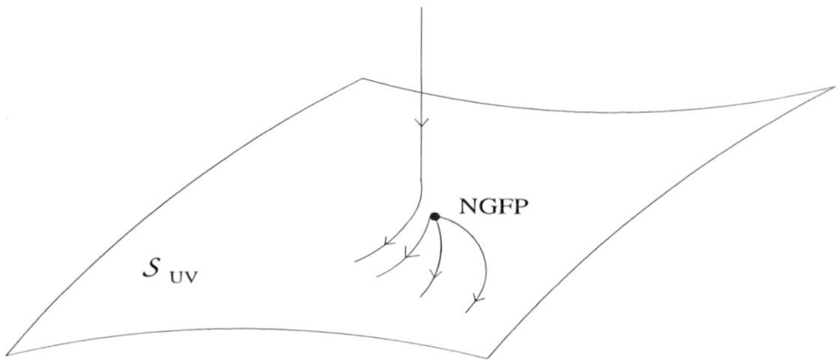

Figure 2. Schematic picture of the UV critical hypersurface $\mathcal{S}_{\rm UV}$ of the NGFP. It is spanned by RG trajectories emanating from the NGFP as the RG scale k is lowered. Trajectories not in the surface are attracted towards $\mathcal{S}_{\rm UV}$ as k decreases. (The arrows point in the direction of decreasing k, from the "UV" to the "IR".)

running of the $\bar{u}$'s is given by

$$\overline{u}_\alpha(k) = k^{d_\alpha}\, u_\alpha(k)\,. \tag{1}$$

Therefore, even directly at a NGFP where $u_\alpha(k) \equiv u_\alpha^*$, the dimensionful couplings keep running according to a power law involving their canonical dimensions d_α:

$$\overline{u}_\alpha(k) = u_\alpha^*\, k^{d_\alpha}\,. \tag{2}$$

Furthermore, non-essential dimensionless couplings are not required to attain fixed point values.

Given a NGFP, an important concept is its *UV critical hypersurface* $\mathcal{S}_{\rm UV}$, or synonymously, its *unstable manifold.* By definition, it consists of all points of theory space which are pulled into the NGFP by the inverse RG flow, i.e., for *in*creasing k. Its dimensionality $\dim\left(\mathcal{S}_{\rm UV}\right) \equiv \Delta_{\rm UV}$ is given by the number of attractive (for *in*creasing cutoff k) directions in the space of couplings.

Writing the RG equations as $k\,\partial_k u_\alpha = \beta_\alpha(u_1, u_2, \ldots)$, the linearized flow near the fixed point is governed by the Jacobi matrix $\mathbf{B} = (B_{\alpha\gamma})$, $B_{\alpha\gamma} \equiv$

$\partial_\gamma \beta_\alpha(u^*)$:

$$k\, \partial_k\, u_\alpha(k) = \sum_\gamma B_{\alpha\gamma}\, \left(u_\gamma(k) - u^*_\gamma\right)\ . \tag{3}$$

The general solution to this equation reads

$$u_\alpha(k) = u^*_\alpha + \sum_I C_I\, V^I_\alpha\, \left(\frac{k_0}{k}\right)^{\theta_I} \tag{4}$$

where the V^I's are the right-eigenvectors of $\mathbf{B}$ with eigenvalues $-\theta_I$, i.e., $\sum_\gamma B_{\alpha\gamma}\, V^I_\gamma = -\theta_I\, V^I_\alpha$. Since $\mathbf{B}$ is not symmetric in general the θ_I's are not guaranteed to be real. We assume that the eigenvectors form a complete system though. Furthermore, k_0 is a fixed reference scale, and the C_I's are constants of integration.

If $u_\alpha(k)$ is to describe a trajectory in $\mathcal{S}_{\text{UV}}$, $u_\alpha(k)$ must approach u^*_α in the limit $k \to \infty$ and therefore we must set $C_I = 0$ for all I with $\text{Re}\,\theta_I < 0$. Hence the dimensionality Δ_{UV} equals the number of $\mathbf{B}$-eigenvalues with a negative real part, i.e., the number of θ_I's with $\text{Re}\,\theta_I > 0$. The corresponding eigenvectors span the tangent space to $\mathcal{S}_{\text{UV}}$ at the NGFP.

If $u_\alpha(k)$ describes a generic trajectory with all C_I nonzero and we *lower* the cutoff, only Δ_{UV} "relevant" parameters corresponding to the eigendirections tangent to $\mathcal{S}_{\text{UV}}$ grow ($\text{Re}\,\theta_I > 0$), while the remaining "irrelevant" couplings pertaining to the eigendirections normal to $\mathcal{S}_{\text{UV}}$ decrease ($\text{Re}\,\theta_I < 0$). Thus near the NGFP a generic trajectory is attracted towards $\mathcal{S}_{\text{UV}}$, see fig. 2.

Coming back to the asymptotic safety construction, let us now use this fixed point in order to take the limit $k \to \infty$. The trajectories which define an infinite cutoff limit for QEG are special in that all irrelevant couplings are set to zero: $C_I = 0$ if $\text{Re}\,\theta_I < 0$. These conditions place the trajectory exactly on $\mathcal{S}_{\text{UV}}$. There is a Δ_{UV}-parameter family of such trajectories, and the experiment must decide which one is realized in Nature. Therefore the predictive power of the theory increases with decreasing dimensionality of $\mathcal{S}_{\text{UV}}$, i.e., number of UV attractive eigendirections of the NGFP. (If $\Delta_{\text{UV}} < \infty$, the quantum field theory thus constructed is comparable to and as predictive as a perturbatively renormalizable model with Δ_{UV} "renormalizable couplings", i.e., couplings relevant at the GFP.)

The quantities θ_I are referred to as critical exponents since when the renormalization group is applied to critical phenomena (second order phase transitions) the traditionally defined critical exponents are related to the θ_I's in a simple way [28]. In fact, one of the early successes of the RG ideas was an explanation of the universality properties of critical phenomena, i.e., the fact that systems at the critical point seem to "forget" the precise form of their microdynamics and just depend on the universality class, characterized by a set of critical exponents, they belong to.

In the present context, "universality" means that certain, very special, quantities related to the RG flow are independent of the precise form of the cutoff and, in particular, its shape function $R^{(0)}$. Universal quantities are potentially measurable or at least closely related to observables. The θ_I's are examples of universal quantities, while the coordinates of the fixed point, u^*_α, are not, even in an exact calculation. Quantities independently known to be universal provide an important tool for testing the reliability or accuracy of *approximate* RG calculations and of truncations in particular. Since they are known to be $R^{(0)}$ independent in an exact treatment, we can determine the degree of their $R^{(0)}$-dependence within the truncation and use it as a measure for the quality of the truncated calculation.

For a more detailed and formal discussion of asymptotic safety and, in particular, its relation to perturbation theory we refer to the review [23].

6. Average Action Approach to Asymptotic Safety

Our discussion of the asymptotic safety construction in the previous section was at the level of the exact (untruncated) RG flow. In this section we are going to implement these ideas in the context of explicitly computable approximate RG flows on truncated theory spaces. We shall mostly concentrate on the Einstein-Hilbert ("R–") and the R^2–truncation of pure gravity in $d = 4$. The corresponding d-dimensional flow equations were derived in refs. [1] and [5], respectively.

6.1. The Phase Portrait of the Einstein-Hilbert Truncation

In [4] the RG equations (14) implied by the Einstein-Hilbert truncation have been analyzed in detail, using both analytical and numerical methods. In partic-

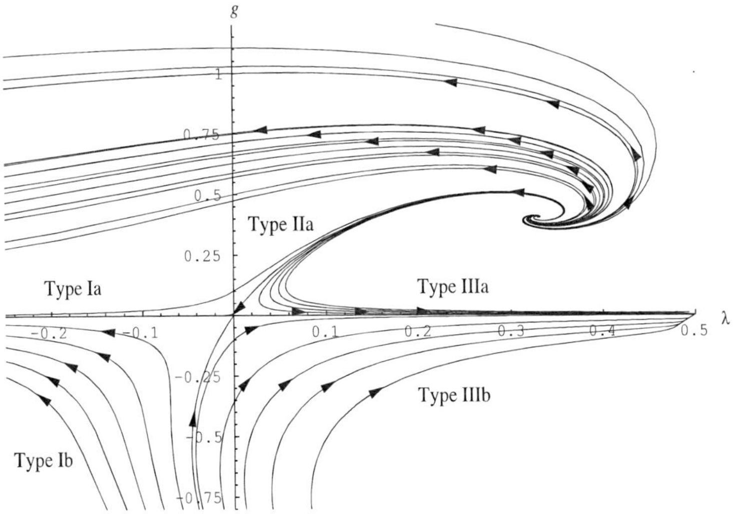

Figure 3. RG flow in the g-λ–plane. The arrows point in the direction of increasing coarse graining, i.e., of decreasing k. (From [4].)

ular all RG trajectories of this system have been classified, and examples have been computed numerically. The most important classes of trajectories in the phase portrait on the g-λ–plane are shown in fig. 3. The trajectories were obtained by numerically solving the system (19) for a sharp cutoff; using a smooth one all qualitative features remain unchanged. The RG flow is found to be dominated by two fixed points (g^*, λ^*): the GFP at $g^* = \lambda^* = 0$, and a NGFP with $g^* > 0$ and $\lambda^* > 0$. There are three classes of trajectories emanating from the NGFP: trajectories of Type Ia and IIIa run towards negative and positive cosmological constants, respectively, and the single trajectory of Type IIa ("separatrix") hits the GFP for $k \to 0$. The high momentum properties of QEG are governed by the NGFP; for $k \to \infty$, in fig. 3 all RG trajectories on the half–plane $g > 0$ run into this point. Note that near the NGFP the dimensionful Newton constant vanishes for $k \to \infty$ according to $G_k \equiv g_k/k^2 \approx g^*/k^2 \to 0$, while the cosmological constant diverges: $\bar{\lambda}_k \equiv \lambda_k k^2 \approx \lambda^* k^2 \to \infty$.

So, the Einstein-Hilbert truncation does indeed predict the existence of a NGFP with exactly the properties needed for the asymptotic safety construction.

Clearly the crucial question to be analyzed now is whether the NGFP found is the projection of a fixed point in the exact theory on the untruncated theory space or whether it is merely the artifact of an insufficient approximation.

6.2. Testing the Einstein-Hilbert Truncation

We mentioned already that the residual $R^{(0)}$-dependence of universal quantities is a measure for the quality of a truncation. This test has been applied to the Einstein-Hilbert truncation in [8, 3]. We shall display the results in the next subsection. In accordance with the general theory the coordinates of the fixed point (g^*, λ^*) are not universal. However, it can be argued that they should give rise to a universal combination, the product $g^*\lambda^*$ which can be measured in principle [3]. While k and, at a fixed value of k, G_k and $\bar{\lambda}_k$ cannot be measured separately, we may invert the function $k \mapsto G_k$ and insert the result $k = k(G)$ into $\bar{\lambda}_k$. This leads to an in principle experimentally testable relationship $\bar{\lambda} = \bar{\lambda}(G)$ between Newton's constant and the cosmological constant. Here $\bar{\lambda}$ and G should be determined in experiments involving similar scales. In the fixed point regime this relationship reads $\bar{\lambda}(G) = g^*\lambda^*/G$. So, even if this is quite difficult in practice, one can determine the product $g^*\lambda^*$ experimentally. As a consequence in any reliable calculation $g^*\lambda^*$ should be approximately $R^{(0)}$ independent.

The ultimate justification of a given truncation consists in checking that if one adds further terms to it, its physical predictions remain robust. The first step towards testing the robustness of the Einstein-Hilbert truncation near the NGFP against the inclusion of other invariants has been taken in refs. [5, 6] where the R^2–truncation of eq. (21) has been analyzed. The corresponding beta functions for the three generalized couplings g, λ and β have been derived, but they are too complicated to be reproduced here. Suffice it to say that on the 3-dimensional (g, λ, β) space, too, a NGFP has been found which generalizes the one from the pure R–calculation. This allows for a comparison of the fixed point results for the R^2– and the Einstein-Hilbert truncation, and for a check of the approximate $R^{(0)}$ independence of universal quantities in the 3-dimensional setting. For the Einstein-Hilbert truncation the universality analysis has been performed for an arbitrary constant gauge parameter α, including the 'physical' value $\alpha = 0$ [3]. Because of its algebraic complexity the R^2–analysis [5] has been carried out in the simpler $\alpha = 1$ gauge.

6.3. Evidence for Asymptotic Safety in the Einstein-Hilbert and R^2-truncation

We now summarize the results concerning the NGFP which were obtained with the R– (items (1)-(5)) and R^2–truncation (items (6)-(9)), respectively [3, 4, 5, 6]. All properties mentioned below are independent pieces of evidence pointing in the direction that QEG is indeed asymptotically safe in four dimensions. Except for point (5) all results refer to $d = 4$.

(1) Universal existence: Both for type A and type B cutoffs the non-Gaussian fixed point exists for all shape functions $R^{(0)}$. (This generalizes earlier results in [8].) It seems impossible to find an admissible cutoff which destroys the fixed point in $d = 4$. This result is highly non-trivial since in higher dimensions ($d \gtrsim 5$) the existence of the NGFP depends on the cutoff chosen [4].

(2) Positive Newton constant: While the position of the fixed point is scheme dependent, all cutoffs yield *positive* values of g^* and λ^*. A negative g^* might have been problematic for stability reasons, but there is no mechanism in the flow equation which would exclude it on general grounds.

(3) Stability: For any cutoff employed the NGFP is found to be UV attractive in both directions of the λ-g–plane. Linearizing the flow equation according to eq. (3) we obtain a pair of complex conjugate critical exponents $\theta_1 = \theta_2^*$ with positive real part θ' and imaginary parts $\pm\theta''$. In terms of $t = \ln(k/k_0)$ the general solution to the linearized flow equations reads

$$\begin{aligned}(\lambda_k, g_k)^{\mathrm{T}} &= (\lambda^*, g^*)^{\mathrm{T}} + 2\Big\{ \big[\mathrm{Re}\, C \, \cos\left(\theta'' \, t\right) + \mathrm{Im}\, C \, \sin\left(\theta'' \, t\right)\big] \, \mathrm{Re}\, V \\ &\quad + \big[\mathrm{Re}\, C \, \sin\left(\theta'' \, t\right) - \mathrm{Im}\, C \, \cos\left(\theta'' \, t\right)\big] \, \mathrm{Im}\, V \Big\} e^{-\theta' t} \, . \end{aligned} \tag{1}$$

with $C \equiv C_1 = (C_2)^*$ an arbitrary complex number and $V \equiv V^1 = (V^2)^*$ the right-eigenvector of $\mathbf{B}$ with eigenvalue $-\theta_1 = -\theta_2^*$. Eq. (3) implies that, due to the positivity of θ', all trajectories hit the fixed point as t is sent to infinity. The non-vanishing imaginary part θ'' has no impact on the stability. However, it influences the shape of the trajectories which spiral into the fixed point for $k \to \infty$. Thus, the fixed point has the stability properties needed in the asymptotic safety scenario.

Solving the full, non-linear flow equations [4] shows that the asymptotic scaling region where the linearization (1) is valid extends from $k = \infty$ down to about $k \approx m_{\rm Pl}$ with the Planck mass defined as $m_{\rm Pl} \equiv G_0^{-1/2}$. Here $m_{\rm Pl}$ plays a role similar to $\Lambda_{\rm QCD}$ in QCD: it marks the lower boundary of the asymptotic scaling region. We set $k_0 \equiv m_{\rm Pl}$ so that the asymptotic scaling regime extends from about $t = 0$ to $t = \infty$.

(4) Scheme- and gauge dependence: Analyzing the cutoff scheme dependence of θ', θ'', and $g^*\lambda^*$ as a measure for the reliability of the truncation, the critical exponents were found to be reasonably constant within about a factor of 2. For $\alpha = 1$ and $\alpha = 0$, for instance, they assume values in the ranges $1.4 \lesssim \theta' \lesssim 1.8$, $2.3 \lesssim \theta'' \lesssim 4$ and $1.7 \lesssim \theta' \lesssim 2.1$, $2.5 \lesssim \theta'' \lesssim 5$, respectively. The universality properties of the product $g^*\lambda^*$ are even more impressive. Despite the rather strong scheme dependence of g^* and λ^* separately, their product has almost no visible s-dependence for not too small values of s. Its value is

$$g^*\lambda^* \approx \begin{cases} 0.12 \text{ for } \alpha = 1 \\ 0.14 \text{ for } \alpha = 0\,. \end{cases} \tag{2}$$

The difference between the "physical" (fixed point) value of the gauge parameter, $\alpha = 0$, and the technically more convenient $\alpha = 1$ are at the level of about 10 to 20 percent.

(5) Higher and lower dimensions: The beta functions implied by the FRGE are continuous functions of the spacetime dimensionality and it is instructive to analyze them for $d \neq 4$. In ref. [1] it has been shown that for $d = 2 + \epsilon$, $|\epsilon| \ll 1$, the FRGE reproduces Weinberg's [22] fixed point for Newton's constant, $g^* = \frac{3}{38}\epsilon$, and also supplies a corresponding fixed point value for the cosmological constant, $\lambda^* = -\frac{3}{38}\Phi_1^1(0)\epsilon$, with the treshold function given in (18). For arbitrary d and a generic cutoff the RG flow is quantitatively similar to the 4-dimensional one for all d smaller than a certain critical dimension $d_{\rm crit}$, above which the existence or non-existence of the NGFP becomes cutoff-dependent. The critical dimension is scheme dependent, but for any admissible cutoff it lies well above $d = 4$. As d approaches $d_{\rm crit}$ from below, the scheme dependence of the universal quantities increases drastically, indicating that the R-truncation becomes insufficient near $d_{\rm crit}$.

In fig. 4 we show the d-dependence of g^*, λ^*, θ', and θ'' for two versions of the sharp cutoff (with $s = 1$ and $s = 30$, respectively) and for the exponential cutoff with $s = 1$. For $2 + \epsilon \le d \le 4$ the scheme dependence of the critical exponents is rather weak; it becomes appreciable only near $d \approx 6$ [4]. Fig. 4 suggests that the Einstein-Hilbert truncation in $d = 4$ performs almost as well as near $d = 2$. Its validity can be extended towards larger dimensionalities by optimizing the shape function [14].

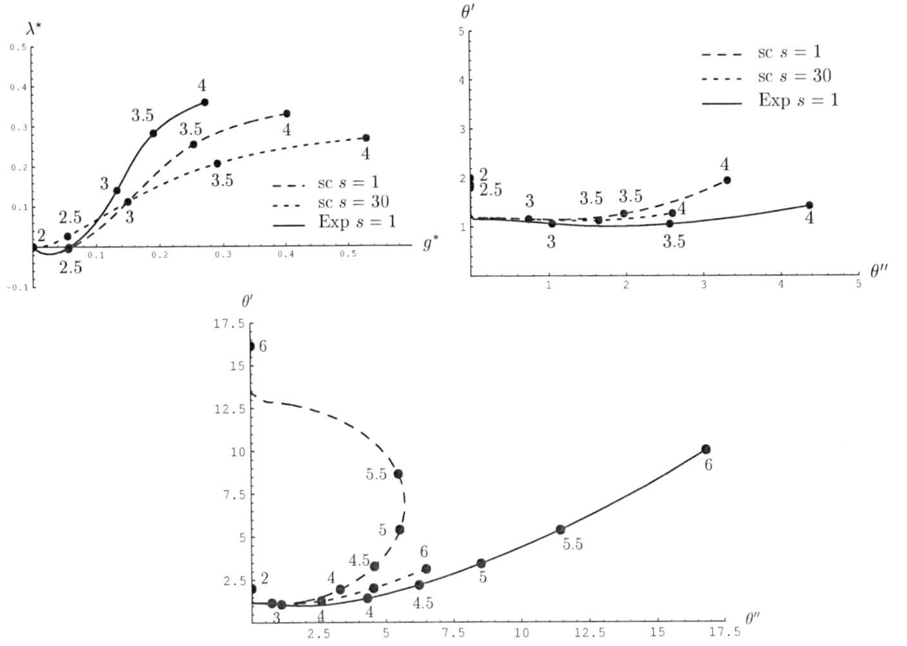

Figure 4. Comparison of λ^*, g^*, θ' and θ'' for different cutoff functions in dependence of the dimension d. Two versions of the sharp cutoff (sc) and the exponential cutoff with $s = 1$ (Exp) have been employed. The upper line shows that for $2 + \epsilon \le d \le 4$ the cutoff scheme dependence of the results is rather small. The lower diagram shows that increasing d beyond about 5 leads to a significant difference in the results for θ', θ'' obtained with the different cutoff schemes. (From [4].)

(6) Position of the fixed point (R^2): Also with the generalized truncation the NGFP is found to exist for all admissible cutoffs. Fig. 5 shows its coordinates $(\lambda^*, g^*, \beta^*)$ for the family of shape functions (20) and the type B cutoff. For every shape parameter s, the values of λ^* and g^* are almost the same as those obtained with the Einstein-Hilbert truncation. In particular, the product $g^*\lambda^*$ is constant with a very high accuracy. For $s = 1$, for instance, one obtains $(\lambda^*, g^*) = (0.348, 0.272)$ from the Einstein-Hilbert truncation and $(\lambda^*, g^*, \beta^*) = (0.330, 0.292, 0.005)$ from the generalized truncation. It is quite remarkable that β^* is always significantly smaller than λ^* and g^*. Within the limited precision of our calculation this means that in the 3-dimensional parameter space the fixed point practically lies on the λ-g–plane with $\beta = 0$, i.e., on the parameter space of the pure Einstein-Hilbert truncation.

(7) Eigenvalues and -vectors (R^2)**:** The NGFP of the R^2-truncation proves to be UV attractive in any of the three directions of the (λ, g, β) space for all cutoffs used. The linearized flow in its vicinity is always governed by a pair of complex conjugate critical exponents $\theta_1 = \theta' + \mathrm{i}\theta'' = \theta_2^*$ with $\theta' > 0$ and a single real, positive critical exponent $\theta_3 > 0$. It may be expressed as

$$(\lambda_k, g_k, \beta_k)^{\mathrm{T}} = (\lambda^*, g^*, \beta^*)^{\mathrm{T}} + 2\Big\{ \big[\mathrm{Re}\, C \, \cos\left(\theta'' t\right) + \mathrm{Im}\, C \, \sin\left(\theta'' t\right)\big] \, \mathrm{Re}\, V + \big[\mathrm{Re}\, C \, \sin\left(\theta'' t\right) - \mathrm{Im}\, C \, \cos\left(\theta'' t\right)\big] \, \mathrm{Im}\, V \Big\} \, e^{-\theta' t} + C_3 V^3 \, e^{-\theta_3 t} \tag{3}$$

with arbitrary complex $C \equiv C_1 = (C_2)^*$ and real C_3, and with $V \equiv V^1 = (V^2)^*$ and V^3 the right-eigenvectors of the stability matrix $(B_{ij})_{i,j \in \{\lambda, g, \beta\}}$ with eigenvalues $-\theta_1 = -\theta_2^*$ and $-\theta_3$, respectively. Clearly the conditions for UV stability are $\theta' > 0$ and $\theta_3 > 0$. They are indeed satisfied for all cutoffs. For the exponential shape function with $s = 1$, for instance, we find $\theta' = 2.15$, $\theta'' = 3.79$, $\theta_3 = 28.8$, and $\mathrm{Re}\, V = (-0.164, 0.753, -0.008)^{\mathrm{T}}$, $\mathrm{Im}\, V = (0.64, 0, -0.01)^{\mathrm{T}}$, $V^3 = -(0.92, 0.39, 0.04)^{\mathrm{T}}$. (The vectors are normalized such that $\|V\| = \|V^3\| = 1$.) The trajectories (6.3) comprise three independent normal modes with amplitudes proportional to $\mathrm{Re}\, C$, $\mathrm{Im} C$ and C_3, respectively. The first two are again of the spiral type while the third one is a straight line.

For any cutoff, the numerical results have several quite remarkable properties. They all indicate that, close to the NGFP, the RG flow is rather well approximated by the pure Einstein-Hilbert truncation.

(a) The β-components of $\mathrm{Re}\, V$ and $\mathrm{Im}\, V$ are very tiny. Hence these two vectors span a plane which virtually coincides with the g-λ–subspace at $\beta = 0$, i.e., with the parameter space of the Einstein-Hilbert truncation. As a conse-

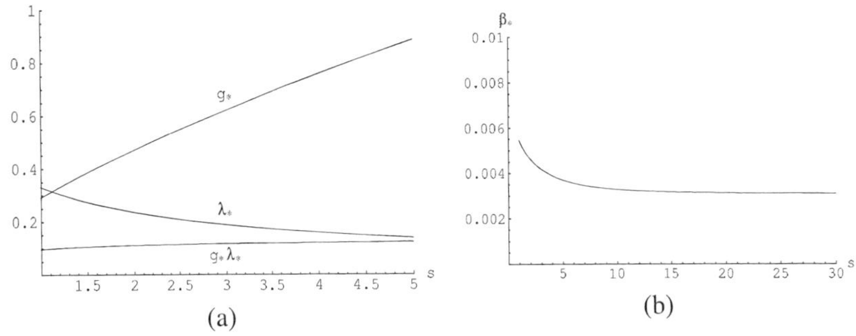

Figure 5. (a) g^*, λ^*, and $g^*\lambda^*$ as functions of s for $1 \leq s \leq 5$, and (b) β^* as a function of s for $1 \leq s \leq 30$, using the family of exponential shape functions (20). (From ref. [6].)

quence, the $\operatorname{Re} C$– and $\operatorname{Im} C$–normal modes are essentially the same trajectories as the "old" normal modes already found without the R^2–term. Also the corresponding θ'– and θ''–values coincide within the scheme dependence.

(b) The new eigenvalue θ_3 introduced by the R^2–term is significantly larger than θ'. When a trajectory approaches the fixed point from below ($t \to \infty$), the "old" normal modes $\propto \operatorname{Re} C, \operatorname{Im} C$ are proportional to $\exp(-\theta' t)$, but the new one is proportional to $\exp(-\theta_3 t)$, so that it decays much quicker. For every trajectory running into the fixed point, i.e., for every set of constants $(\operatorname{Re} C, \operatorname{Im} C, C_3)$, we find therefore that once t is sufficiently large the trajectory lies entirely in the $\operatorname{Re} V$-$\operatorname{Im} V$–subspace, i.e., the $\beta = 0$-plane practically.

Due to the large value of θ_3, the new scaling field is very "relevant". However, when we start at the fixed point ($t = \infty$) and lower t it is only at the low energy scale $k \approx m_{\text{Pl}}$ ($t \approx 0$) that $\exp(-\theta_3 t)$ reaches unity, and only then, i.e., far away from the fixed point, the new scaling field starts growing rapidly.

(c) Since the matrix $\mathbf{B}$ is not symmetric its eigenvectors have no reason to be orthogonal. In fact, one finds that V^3 lies almost in the $\operatorname{Re} V$-$\operatorname{Im} V$–plane. For the angles between the eigenvectors given above we obtain $\sphericalangle(\operatorname{Re} V, \operatorname{Im} V) = 102.3°$, $\sphericalangle(\operatorname{Re} V, V^3) = 100.7°$, $\sphericalangle(\operatorname{Im} V, V^3) = 156.7°$. Their sum is $359.7°$ which confirms that $\operatorname{Re} V$, $\operatorname{Im} V$ and V^3 are almost coplanar. This implies that when we lower t and move away from the fixed point so that the V^3–scaling field starts growing, it is again predominantly the $\int d^dx\,\sqrt{g}$ and $\int d^dx\,\sqrt{g}R$

invariants which get excited, but not $\int d^dx \sqrt{g}R^2$ in the first place.

Summarizing the three points above, we can say that very close to the fixed point the RG flow seems to be essentially two-dimensional, and that this two-dimensional flow is well approximated by the RG equations of the Einstein-Hilbert truncation. In fig. 6 we show a typical trajectory which has all three normal modes excited with equal strength ($\mathrm{Re}\, C = \mathrm{Im}\, C = 1/\sqrt{2}$, $C_3 = 1$). All the way down from $k = \infty$ to about $k = m_{\mathrm{Pl}}$ it is confined to a very thin box surrounding the $\beta = 0$–plane.

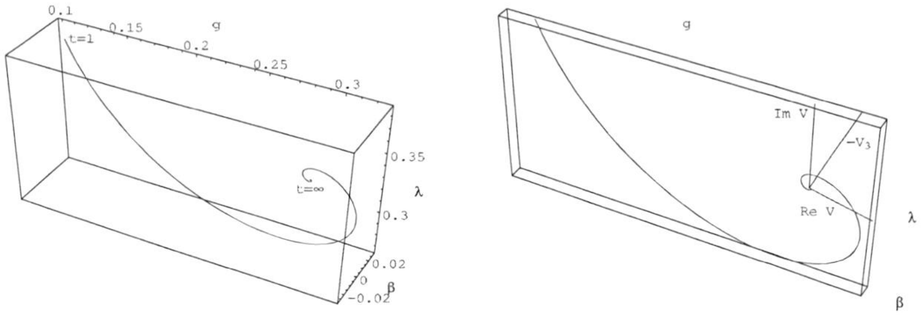

(a) (b)

Figure 6. Trajectory of the linearized flow equation obtained from the R^2–truncation for $1 \leq t = \ln(k/k_0) < \infty$. In (b) we depict the eigendirections and the "box" to which the trajectory is confined. (From ref. [6].)

(8) Scheme Dependence (R^2): The scheme dependence of the critical exponents and of the product $g^*\lambda^*$ turns out to be of the same order of magnitude as in the case of the Einstein-Hilbert truncation. Fig. 7 shows the cutoff dependence of the critical exponents, using the family of shape functions (20). For the cutoffs employed θ' and θ'' assume values in the ranges $2.1 \lesssim \theta' \lesssim 3.4$ and $3.1 \lesssim \theta'' \lesssim 4.3$, respectively. While the scheme dependence of θ'' is weaker than in the case of the Einstein-Hilbert truncation one finds that it is slightly larger for θ'. The exponent θ_3 suffers from relatively strong variations as the cutoff is changed, $8.4 \lesssim \theta_3 \lesssim 28.8$, but it is always significantly larger than θ'. The product $g^*\lambda^*$ again exhibits an extremely weak scheme dependence. Fig. 5(a) displays $g^*\lambda^*$ as a function of s. It is impressive to see how the cutoff dependences of g^* and λ^* cancel almost perfectly. Fig. 5(a) suggests the universal value $g^*\lambda^* \approx 0.14$. Comparing this value to those obtained from the Einstein-

Hilbert truncation we find that it differs slightly from the one based upon the same gauge $\alpha = 1$. The deviation is of the same size as the difference between the $\alpha = 0$– and the $\alpha = 1$–results of the Einstein-Hilbert truncation.

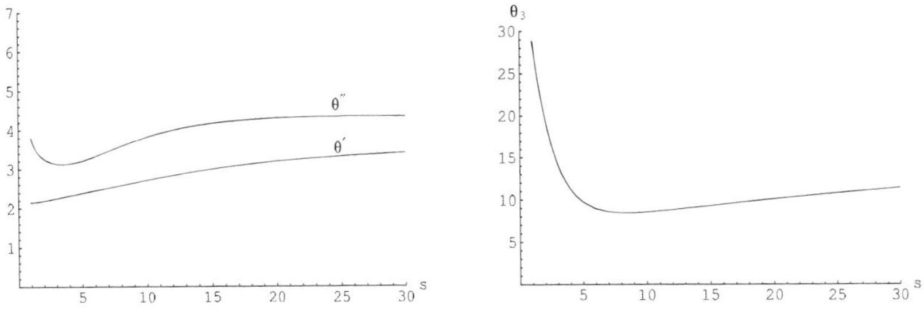

(a) (b)

Figure 7. (a) $\theta' = \mathrm{Re}\,\theta_1$ and $\theta'' = \mathrm{Im}\,\theta_1$, and (b) θ_3 as functions of s, using the family of exponential shape functions (20). (From [5].)

As for the universality of the critical exponents we emphasize that the qualitative properties listed above (e.g., $\theta', \theta_3 > 0$, $\theta_3 \gg \theta'$, etc.) are obtained for all cutoffs. The θ's have a much stronger scheme dependence than $g^*\lambda^*$, however. This is most probably due to neglecting further relevant operators in the truncation so that the B-matrix we are diagonalizing is too small still.

(9) Dimensionality of $\mathcal{S}_{\mathrm{UV}}$: According to the canonical dimensional analysis, the (curvature)^n-invariants in 4 dimensions are classically marginal for $n = 2$ and irrelevant for $n > 2$. The results for θ_3 indicate that there are large non-classical contributions so that there might be relevant operators perhaps even beyond $n = 2$. With the present truncation it is clearly not possible to determine their number Δ_{UV}. However, as it is hardly conceivable that the quantum effects change the signs of arbitrarily large (negative) classical scaling dimensions, Δ_{UV} should be finite [22].

Summary: The above results strongly suggest that the non-Gaussian fixed point occurring in the Einstein-Hilbert truncation is not a truncation artifact but rather the projection of a fixed point in the exact theory space. The fixed point and all its qualitative properties are stable against variations of the cutoff and the inclusion of a further invariant in the truncation. It is particularly remarkable that within the scheme dependence the additional R^2–term has essentially no impact on the fixed point. We interpret the above results and their mutual consistency as quite non-trivial indications supporting the conjecture that 4-dimensional QEG indeed possesses a RG fixed point with precisely the properties needed for its non-perturbative renormalizability and asymptotic safety.

In the next subsection we will see that this general picture remains intact when we go to an even more general truncation.

6.4. The $f(R)$-truncation

The extension of the truncated theory space to arbitrary functions of the Ricci scalar has been carried out in [16, 17, 18]. In this truncation ansatz $S_{\rm gf}$ and $S_{\rm gh}$ are taken to be classical while, in the language of eq. (1),

$$\bar{\Gamma}_k[g] = \int d^dx \sqrt{g}\, f_k(R)\,, \quad \widehat{\Gamma}_k[g,\bar{g}] = 0\,. \tag{4}$$

The Einstein-Hilbert truncation discussed in section 4 corresponds to setting $f_k(R) = (16\pi G_k)^{-1}\,(-R + 2\Lambda_k)$ and using (6) rather than setting $\widehat{\Gamma}_k[g,\bar{g}] = 0$.

Table 1. Location of the NGFP obtained from the partial differential equation (6) by expanding $\mathcal{F}_k(\rho)$ in a power series in ρ up to order ρ^N, including k-dependent coupling constants $u_n(k)$, $n = 0,\ldots,N$.

N	u_0^*	u_1^*	u_2^*	u_3^*	u_4^*	u_5^*	u_6^*	$g^*\lambda^*$
1	0.00523	−0.0202						0.127
2	0.00333	−0.0125	0.00149					0.211
3	0.00518	−0.0196	0.00070	−0.0104				0.134
4	0.00505	−0.0206	0.00026	−0.0120	−0.0101			0.118
5	0.00506	−0.0206	0.00023	−0.0105	−0.0096	−0.00455		0.119
6	0.00504	−0.0208	0.00012	−0.0110	−0.0109	−0.00473	0.00238	0.116

Remarkably, combining the transverse-traceless decomposition of the quantum metric (22) with the "anharmonic" gauge choice $\varpi = 0$, $\alpha = 0$ in (4) and the optimized cutoff shape function of ref. [49], the projection of the traces appearing in (21) onto the truncation subspace can be carried out explicitly, requiring knowledge about the early time expansion of the heat-kernel up to fourth

Table 2. Stability coefficients of the NGFP for increasing dimension $N+1$ of the truncation subspace. The first two critical exponents are a complex pair $\theta_\pm = \theta' \pm i\theta''$.

N	θ'	θ''	θ_2	θ_3	θ_4	θ_5	θ_6
1	2.38	-2.17					
2	1.26	-2.44	27.0				
3	2.67	-2.26	2.07	-4.42			
4	2.83	-2.42	1.54	-4.28	-5.09		
5	2.57	-2.67	1.73	-4.40	$-3.97+4.57i$	$-3.97-4.57i$	
6	2.39	-2.38	1.51	-4.16	$-4.67+6.08i$	$-4.67-6.08i$	-8.67

order in the curvature only [16, 17]. Introducing the dimensionless quantities

$$\rho \equiv k^{-2}R\,, \qquad \mathcal{F}_k(\rho) \equiv k^{-4} f_k(R/k^2) \tag{5}$$

the autonomous partial differential equation governing the scale-dependence of $\mathcal{F}_k$ can be obtained [17]

$$\begin{aligned}
&384\pi^2\left(\partial_t \mathcal{F}_k + 4\mathcal{F}_k - 2\rho\mathcal{F}_k'\right)\\
&= \left[5\rho^2\theta\left(1-\tfrac{\rho}{3}\right) - \left(12 + 4\rho - \tfrac{61}{90}\rho^2\right)\right]\left[1-\tfrac{\rho}{3}\right]^{-1} + 10\,\rho^2\,\theta\left(1-\tfrac{\rho}{3}\right)\\
&+ \left[10\,\rho^2\,\theta(1-\tfrac{\rho}{4}) - \rho^2\,\theta(1+\tfrac{\rho}{4}) - \left(36 + 6\,\rho - \tfrac{67}{60}\rho^2\right)\right]\left[1-\tfrac{\rho}{4}\right]^{-1}\\
&+ \left[\eta_f\,\left(10 - 5\rho - \tfrac{271}{36}\rho^2 + \tfrac{7249}{4536}\rho^3\right) + \left(60 - 20\rho - \tfrac{271}{18}\rho^2\right)\right]\left[1 + \tfrac{\mathcal{F}_k}{\mathcal{F}_k'} - \tfrac{\rho}{3}\right]^{-1}\\
&+ \frac{5\rho^2}{2}\left[\eta_f\left((1+\tfrac{\rho}{3})\theta(1+\tfrac{\rho}{3}) + (2+\tfrac{\rho}{3})\theta(1+\tfrac{\rho}{6})\right) + 2\theta(1+\tfrac{\rho}{3}) + 4\theta(1+\tfrac{\rho}{6})\right]\\
&\quad\times\left[1 + \tfrac{\mathcal{F}_k}{\mathcal{F}_k'} - \tfrac{\rho}{3}\right]^{-1}\\
&+ \Big[\mathcal{F}_k'\,\eta_f\,\left(6 + 3\rho + \tfrac{29}{60}\rho^2 + \tfrac{37}{1512}\rho^3\right)\\
&\quad + \left(\partial_t\mathcal{F}_k'' - 2\rho\mathcal{F}_k'''\right)\left(27 - \tfrac{91}{20}\rho^2 - \tfrac{29}{30}\rho^3 - \tfrac{181}{3360}\rho^4\right)\\
&\quad + \mathcal{F}_k''\left(216 - \tfrac{91}{5}\rho^2 - \tfrac{29}{15}\rho^3\right) + \mathcal{F}_k'\left(36 + 12\rho + \tfrac{29}{30}\rho^2\right)\Big]\\
&\quad\times\left[2\mathcal{F}_k + 3\mathcal{F}_k'(1-\tfrac{2}{3}\rho) + 9\mathcal{F}_k''(1-\tfrac{\rho}{3})^2\right]^{-1}.
\end{aligned} \tag{6}$$

Here $\eta_f \equiv (\mathcal{F}_k')^{-1}\,(\partial_t \mathcal{F}_k' + 2\mathcal{F}_k' - 2\rho\mathcal{F}_k'')$ provides the natural generalization of (minus) the anomalous dimension of Newton's constant η_N.

Based on this equation, the search for the NGFP on truncation subspaces involving higher powers of the curvature scalar reduces to an algebraic problem. In this course we make a polynomial ansatz for $\mathcal{F}_k(\rho)$, setting

$$\mathcal{F}_k(\rho) = \sum_{n=0}^{N} u_n(t)\,\rho^n\,, \qquad N \in \mathbb{N}\,. \tag{7}$$

Here, $u_1(k) = -(16\pi g_k)^{-1}$, so that a negative $u_1(k)$ actually corresponds to a positive Newton's constant. Substituting this ansatz into (6), expanding its RHS in powers of ρ, and matching the coefficients up to ρ^N yields the non-perturbative β-functions for the dimensionless couplings $u_n(t)$,

$$\partial_t u_n = \beta_{u_n}(u_0, \ldots, u_N)\,, \quad n = 0, \ldots, N\,. \tag{8}$$

The resulting β-functions indeed give rise to a NGFP with $g^* > 0, \lambda^* > 0$, whose N-dependent position is shown in Table 1. In particular the product $g^*\lambda^* = u_0/(32\pi(u_1)^2)$ displayed in the last column is remarkably constant. It is in excellent agreement with the Einstein-Hilbert truncation (2), and the R^2-truncation, fig. 5(a). Only the value obtained in the case $N = 2$ shows a mild deviation from these earlier computations, which can, most probably, be attributed to the use of a different gauge-fixing procedure, cutoff shape function, and ansatz for $\widehat{\Gamma}_k$.

Performing the stability analysis of the RG flow around the NGFP as explained in section 5 we also obtain the stability coefficients of the NGFP. These are summarized in Table 2. In particular we find that only three of the eigendirections associated to the NGFP are relevant, i.e., UV attractive. Including higher derivative terms $R^n, n \geq 3$, in the truncation creates irrelevant directions only. Thus, in this truncation $\Delta_{\rm UV} = 3$, i.e., the UV critical surface associated to the fixed point is a 3-dimensional submanifold in the truncated theory space. It is stable with respect to increasing N. The RG trajectories tracing out this surface are determined by fixing the three relevant couplings. These describe in which direction tangent to $\mathcal{S}_{\rm UV}$ the trajectory flows away from the fixed point. All remaining couplings, the irrelevant ones, are predictions from asymptotic safety. In [18] these results have been extended up to $N = 8$, providing even stronger evidence for the robustness of the RG flow under the inclusion of further invariants.

Summary: Truncations involving higher-order polynomials in R fully confirm the general picture suggested by the simple Einstein-Hilbert truncation. The corresponding quantum field theories are found to have only three free parameters.

7. Conclusion

Before closing, some further comments might be helpful here.

(1) The construction of an effective average action for gravity as introduced in [1] represents a *"background independent"* approach to quantum gravity. Somewhat paradoxically, this background independence is achieved by means of the background field formalism: One fixes an arbitrary background, quantizes the fluctuation field in this background, and afterwards adjusts $\bar{g}_{\mu\nu}$ in such a way that the expectation value of the fluctuation vanishes: $\bar{h}_{\mu\nu} = 0$. In this way the background gets fixed dynamically. The results of refs. [19, 20] suggest that a proper implementation of "background independence" might be crucial for asymptotic safety.

(2) The combination of the effective average action with the background field method has been successfully tested within conventional field theory. In QED and Yang-Mills type gauge theories it reproduces the known results and extends them into the non-perturbative domain [27, 29].

(3) The coexistence of asymptotic safety and perturbative non-renormalizability is well understood. In particular upon fixing $\bar{g}_{\mu\nu} = \eta_{\mu\nu}$ and expanding the trace on its RHS in powers of G the FRGE reproduces the divergences of perturbation theory; see ref. [23] for a detailed discussion of this point.

(4) It is to be emphasized that in the average action framework the RG flow, i.e., the vector field $\vec{\beta}$, is completely determined once a theory space is fixed. As a consequence, the choice of theory space determines the set of fixed points Γ^* at which asymptotically safe theories can be defined. Therefore, in the asymptotic safety scenario the bare action $S = \Gamma^*$ is a *prediction* of the theory rather than an ad hoc postulate as usually in quantum field theory. (Ambiguities could arise only if there is more than one suitable NGFP.)

(5) According to the results available to date, the Einstein-Hilbert action of classical General Relativity seems not to play any distinguished role in the asymptotic safety context, at least not at the conceptual level. The only known NGFP on the theory space of QEG has the structure $\Gamma^* =$ Einstein-Hilbert action + "more" where "more" stands for both local and non-local corrections. So it seems that the Einstein-Hilbert action is only an approximation to the true fixed point action, albeit an approximation which was found to be rather reliable for many purposes.

(6) Any quantum theory of gravity must reproduce the successes of classical General Relativity. As for QEG, it cannot be expected that this will happen for all RG trajectories in $\mathcal{S}_{\mathrm{UV}}$, but it should happen for some or at least one of them. Within the Einstein-Hilbert truncation it has been shown [46] that there actually do exist trajectories (of Type IIIa) which have an extended classical regime and are consistent with all observations.

(7) In the classical regime mentioned above the spacetime geometry is non-dynamical to a very good approximation. In this regime the familiar methods of quantum field theory in curved classical spacetimes apply, and it is clear therefore that effects such as Hawking radiation or cosmological particle production are reproduced by the general framework of QEG with matter.

(8) Coupling free massless matter fields to gravity, it turned out [12] that the fixed point continues to exist under very weak conditions concerning the number of various types of matter fields (scalars, fermions, etc.). No fine tuning with respect to the matter multiplets is necessary. In particular asymptotic safety does not seem to require any special constraints or symmetries among the matter fields such as supersymmetry, for instance.

(9) Since the NGFP seems to exist already in pure gravity it is likely that a widespread prejudice about gravity may be incorrect: its quantization seems not to require any kind of unification with the other fundamental interactions.

Given the situation that by now the asymptotic safety of QEG hardly can be questioned any more, future work will have to focus on its physics implications. The effective average action is an ideal framework for investigations of this sort since, contrary to other exact RG schemes, it provides a family of scale dependent *effective* (rather than *bare*) actions, $\{\Gamma_k[\,\cdot\,], 0 \leq k < \infty\}$. Dealing with phenomena involving typical scales k, a tree–level evaluation of Γ_k is sufficient for finding the leading quantum gravity effects. The investigations already performed in this direction employed the following methods.

(a) *RG improvement*: In refs. [38] and [40], respectively, a first study of the asymptotic safety-based "phenomenology" of black hole and cosmological spacetimes has been carried out by "RG improving" the classical field equations or their solutions. Hereby k is identified with a fixed, geometrically motivated scale. Using the same method, modified dispersion relations of point particles were discussed in [47].

(b) *Scale dependent geometry:* In the spirit of the gravitational average action, a spacetime manifold can be visualized as a fixed differentiable manifold equipped with infinitely many metric structures $\{\langle g_{\mu\nu}\rangle_k, 0 \leq k < \infty\}$ where $\langle g_{\mu\nu}\rangle_k$ is a solution to the effective field equation implied by Γ_k. Comparable to the situation in fractal geometry the metric, and therefore all distances, depend on the resolution of the experiment by means of which spacetime is probed. A general discussion of the geometrical issues involved (scale dependent diffeomorphisms, symmetries, causal structures, etc.) was given in [32], and in [31] these ideas were applied to show that QEG can generate a minimum length dynamically. In [3, 5] it has been pointed out that the QEG spacetimes should have fractal properties, with a fractal dimension equal to 4 on macroscopic and 2 on microscopic scales. This picture was confirmed by the computation of their spectral dimension in [33]. Quite remarkably, the same dynamical dimensional reduction from 4 to 2 has also been observed in Monte-Carlo simulations using the causal triangulation approach [34, 35, 36]. It is therefore intriguing to speculate that this discrete approach and the gravitational average action actually describe the same underlying theory.

Acknowledgments

F.S. acknowledges financial support from the ANR grant BLAN06-3-137168.

References

[1] M. Reuter, *Phys. Rev. D* 57 971 (1998), arXiv:hep-th/9605030.

[2] D. Dou and R. Percacci, *Class. Quant. Grav.* 15 3449 (1998).

[3] O. Lauscher and M. Reuter, *Phys. Rev. D* 65 025013 (2002), arXiv:hep-th/0108040.

[4] M. Reuter and F. Saueressig, *Phys. Rev. D* 65 065016 (2002), arXiv:hep-th/0110054.

[5] O. Lauscher and M. Reuter, *Phys. Rev. D*66 025026 (2002), arXiv:hep-th/0205062.

[6] O. Lauscher and M. Reuter, *Class. Quant. Grav.* 19 483 (2002), arXiv:hep-th/0110021.

[7] O. Lauscher and M. Reuter, *Int. J. Mod. Phys. A* 17 993 (2002), arXiv:hep-th/0112089.

[8] W. Souma, *Prog. Theor. Phys.* 102 181 (1999).

[9] M. Reuter and F. Saueressig, *Phys. Rev. D* 66 125001 (2002), arXiv:hep-th/0206145; *Fortschr. Phys.* 52 650 (2004), arXiv:hep-th/0311056.

[10] A. Bonanno, M. Reuter, *JHEP* 02 035 (2005), arXiv:hep-th/0410191.

[11] For reviews on QEG see:
M. Reuter and F. Saueressig, arXiv:0708.1317 [hep-th];
O. Lauscher and M. Reuter in *Quantum Gravity*, B. Fauser, J. Tolksdorf and E. Zeidler (Eds.), Birkhäuser, Basel (2007) and arXiv:hep-th/0511260;
O. Lauscher and M. Reuter in *Approaches to Fundamental Physics*, I.-O. Stamatescu and E. Seiler (Eds.), Springer, Berlin (2007).

[12] R. Percacci and D. Perini, Phys. Rev. D 67 081503 (2003); Phys. Rev. D 68 044018 (2003); D. Perini, *Nucl. Phys. Proc. Suppl.* 127 C 185 (2004).

[13] A. Codello and R. Percacci, *Phys. Rev. Lett.* 97 221301 (2006).

[14] D. Litim, *Phys. Rev. Lett.* 92 201301 (2004); *AIP Conf. Proc.* 841 322 (2006).
P. Fischer and D. Litim, *Phys. Lett. B* 638 (2006) 497; *AIP Conf. Proc.* 861 336 (2006).

[15] R. Percacci and D. Perini, Class. Quant. Grav. 21 5035 (2004).

[16] A. Codello, R. Percacci and C. Rahmede, *Int. J. Mod. Phys. A* 23 143 (2008).

[17] P.F. Machado and F. Saueressig, *Phys. Rev. D* 77 124045 (2008), arXiv:0712.0445 [hep-th].

[18] A. Codello, R. Percacci and C. Rahmede, preprint arXiv:0805.2909 [hep-th].

[19] M. Reuter and H. Weyer, preprint arXiv:0801.3287 [hep-th].

[20] M. Reuter and H. Weyer, preprint arXiv:0804.1475 [hep-th].

[21] J. E. Daum and M. Reuter, preprint arXiv:0806.3907 [hep-th].

[22] S. Weinberg in *General Relativity, an Einstein Centenary Survey*, S.W. Hawking and W. Israel (Eds.), Cambridge University Press, 1979; S. Weinberg, arXiv:hep-th/9702027.

[23] For comprehensive reviews of asymptotic safety in gravity see:
M. Niedermaier and M. Reuter, *Living Reviews in Relativity* 9 5 (2006);
R. Percacci, preprint arXiv:0709.3851 [hep-th].

[24] For general introductions and reviews see:
C. Kiefer, *Quantum Gravity*, Second Edition, Oxford Science Publications, Oxford, 2007;
C. Rovelli, *Quantum Gravity*, Cambridge University Press, Cambridge, 2004;
A. Ashtekar, *Lectures on non-perturbative canonical gravity*, World Scientific, Singapore (1991);
A. Ashtekar and J. Lewandowski, *Class. Quant. Grav.* 21 R53 (2004);
T. Thiemann, *Modern Canonical Quantum General Relativity*, Cambridge University Press, Cambridge (2007).

[25] P. Forgács and M. Niedermaier, arXiv:hep-th/0207028;
M. Niedermaier, *JHEP* 12 066 (2002); *Nucl. Phys. B* 673 131 (2003); *Class. Quant. Grav.* 24 R171 (2007).

[26] C. Wetterich, *Phys. Lett. B* 301 90 (1993).

[27] M. Reuter and C. Wetterich, *Nucl. Phys. B* 417 181 (1994), *Nucl. Phys. B* 427 291 (1994), *Nucl. Phys. B* 391 147 (1993), *Nucl. Phys. B* 408 91

(1993); M. Reuter, Phys. Rev. D **53** 4430 (1996), *Mod. Phys. Lett. A* 12 2777 (1997).

[28] For a review see: J. Berges, N. Tetradis and C. Wetterich, *Phys. Rep.* 363 223 (2002); C. Wetterich, *Int. J. Mod. Phys. A* 16 1951 (2001).

[29] For reviews of the effective average action in Yang-Mills theory see: M. Reuter, arXiv:hep-th/9602012; J. Pawlowski, arXiv:hep-th/0512261; H. Gies, arXiv:hep-ph/0611146.

[30] For reviews see: C. Bagnuls and C. Bervillier, *Phys. Rep.* 348 91 (2001); T.R. Morris, *Prog. Theor. Phys. Suppl.* 131 395 (1998); J. Polonyi, *Central Eur. J. Phys.* 1 1 (2004).

[31] M. Reuter and J. Schwindt, *JHEP* 01 070 (2006), arXiv:hep-th/0511021.

[32] M. Reuter and J. Schwindt, *JHEP* 01 049 (2007), arXiv:hep-th/0611294.

[33] O. Lauscher and M. Reuter, *JHEP* 10 050 (2005), arXiv:hep-th/0508202.

[34] J. Ambjørn, J. Jurkiewicz and R. Loll, *Phys. Rev. Lett.* 93 131301 (2004).

[35] J. Ambjørn, J. Jurkiewicz and R. Loll, *Phys. Lett. B* 607 205 (2005).

[36] J. Ambjørn, J. Jurkiewicz and R. Loll, *Phys. Rev. Lett.* 95 171301 (2005); *Phys. Rev. D* 72 064014 (2005); *Contemp. Phys.* 47 103 (2006).

[37] A. Connes, *JHEP* 11 081 (2006); A.H. Chamseddine, A. Connes and M. Marcolli, arXiv:hep-th/0610241.

[38] A. Bonanno and M. Reuter, *Phys. Rev. D* 62 043008 (2000), arXiv:hep-th/0002196; *Phys. Rev. D* 73 083005 (2006), arXiv:hep-th/0602159; *Phys. Rev. D* 60 084011 (1999), arXiv:gr-qc/9811026.

[39] M. Reuter and E. Tuiran, arXiv:hep-th/0612037.

[40] A. Bonanno and M. Reuter, Phys. Rev. D 65 043508 (2002),arXiv:hep-th/0106133;
M. Reuter and F. Saueressig, *JCAP* 09 012 (2005), arXiv:hep-th/0507167.

[41] A. Bonanno and M. Reuter, *Phys. Lett. B* 527 9 (2002), arXiv:astro-ph/0106468;
Int. J. Mod. Phys. D 13 107 (2004), arXiv:astro-ph/0210472.

[42] E. Bentivegna, A. Bonanno and M. Reuter, *JCAP* 01 001 (2004), arXiv:astro-ph/0303150.

[43] A. Bonanno, G. Esposito and C. Rubano, *Gen. Rel. Grav.* 35 1899 (2003); *Class. Quant. Grav.* 21 5005 (2004); A. Bonanno, G. Esposito, C. Rubano and P. Scudellaro, *Class. Quant. Grav.* 23 3103 (2006) and 24 1443 (2007).

[44] M. Reuter and H. Weyer, *Phys. Rev. D* 69 104022 (2004), arXiv:hep-th/0311196.

[45] M. Reuter and H. Weyer, *Phys. Rev. D* 70 124028 (2004), arXiv:hep-th/0410117.

[46] M. Reuter and H. Weyer, *JCAP* 12 001 (2004), arXiv: hep-th/0410119.

[47] F. Girelli, S. Liberati, R. Percacci, C. Rahmede, arXiv:gr-qc/0607030.

[48] J. Moffat, *JCAP* 05 003 (2005), astro-ph/0412195; J.R. Brownstein and J. Moffat, Astrophys. J. 636 721 (2006); *Mon. Not. Roy. Astron. Soc.* 367 527 (2006).

[49] D. Litim, Phys. Lett. B 486 92 (2000); *Phys. Rev. D* 64 105007 (2001); *Int. J. Mod. Phys. A* 16 2081 (2001).

[50] M. Reuter and C. Wetterich, *Nucl. Phys. B* 506 483 (1997) , arXiv:hep-th/9605039.

[51] L.F. Abbott, *Nucl. Phys. B* 185 189 (1981); B.S. DeWitt, *Phys. Rev.* 162 1195 (1967); M.T. Grisaru, P. van Nieuwenhuizen and C.C. Wu, *Phys. Rev. D* 12 3203 (1975); D.M. Capper, J.J. Dulwich and M. Ramon Medrano, *Nucl. Phys. B* 254 737 (1985); S.L. Adler, *Rev. Mod. Phys.* 54 729 (1982).

[52] For an introduction see: M. Böhm, A. Denner, H. Joos, *Gauge Theories of the Strong and Electroweak Interactions*, Teubner, Stuttgart, 2001.

[53] D. Litim and J. Pawlowski, *Phys. Lett. B* 435 181 (1998).

[54] S. Falkenberg and S.D. Odintsov, *Int. J. Mod. Phys. A* 13 607 (1998).

[55] L.N. Granda, *Europhys. Lett.* 42 487 (1998).

[56] G.O. Pires, *Int. J. Mod. Phys. A* 13 5425 (1998).

[57] R. Percacci, preprint arXiv:hep-th/0409199.

In: Recent Research in Quantum Gravity ISBN: 978-1-61942-386-2
Editor: A. Dasgupta, pp. 95-114

Chapter 4

MIDI-SUPERSPACE MODELS FOR QUANTUM BLACK HOLES

***J. Gegenberg*[1] *and G. Kunstatter*[2]**
[1]Department of Mathematics and Statistics and
Department of Physics,
University of New Brunswick,
Fredericton, New Brunswick, Canada
[2]Department of Physics and
Winnipeg Institute for Theoretical Physics,
University of Winnipeg,
Winnipeg, Manitoba, Canada

Abstract

We describe some recent results that have emerged in the context of generic dilaton gravity (GDG) in two space-time dimensions. First we review the Hamiltonian analysis highlighting the definition and properties of the generalized Misner–Sharpe mass function, which plays an important role in the study of black hole formation, dynamical horizons and singularity resolution. We then show how a partial gauge fixing leads generically to a local Hamiltonian constraint free of spatial derivatives. Finally we review the results obtained from the polymer quantization of the Hamiltonian constraint in the vacuum theory, with emphasis on the discrete spatial structure that naturally emerges.

1. Introduction

There are several fundamental problems associated with the postulated existence of black holes. These include the inevitability of classical singularities, the information loss paradox and the microscopic source of the Bekenstein–Hawking entropy. The resolution of any of these issues will almost certainly consitute a major step towards understanding quantum gravity. In fact, the main theoretical success of both string theory and loop quantum gravity is the calculation from microscopic quantities of the Bekenstein–Hawking entropy for certain types of black holes.

Midi-superspace models by definition are field theories that retain many essential properties of the original theory of physical interest while being sufficiently simple so as to be tractable. In the context of black hole physics, a particularly desirable attribute for midi-superspace models is diffeomorphism invariance. It is likely that any viable resolution of the black hole puzzles alluded to above will be generic in the senses that it will be the same irrespective of which gravitational action is valid in the classical limit and which underlying microscopic theory of quantum gravity turns out to be correct. For these reasons, spherically symmetric gravity [1], and its generalization generic 2-d dilaton gravity (GDG) [2–4] provide ideal theoretical laboratories for studying these issues. Not only are they solvable and invariant under two dimensional diffeormorphisms, but they comprise a large class of theories that can be studied simultaneously within a single formalism.

Significant support for the view that GDG contains sufficient information to be useful in the study of quantum black holes comes from recent work of Carlip [5, 6], who argued that the microscopic degrees of freedom that account for black hole entropy may be found in gauge modes, more accurately diffeomorphisms, that become physical due to the imposition of horizon boundary conditions. Remarkably, he was able to provide solid, if not definitive, arguments for this viewpoint in the context of generic dilaton gravity. One version [5] started from the Hamiltonian analysis of Lorentzian GDG, but ran into technical problems related to the need to specify horizon boundary conditions along a null surface. This required the use of a "stretched" horizon that was spacelike. A more complete and rigorous argument was constructed [6] in the context of Euclidean black holes in GDG, in which the horizon is located at a point instead of along a null surface. The imposition of a horizon constraint modified the structure of the diffeomorphism constraint algebra so as to make some of the would be gauge degrees of freedom physical. The result was an infinite number of degrees that formed a conformal field theory. This allowed the use of standard CFT techniques to compute the corresponding density of states. Remarkably, the statistical mechnical entropy associated with these would be gauge degrees of freedom was equal to the standard Bekenstein–Hawking entropy for the large class of black hole spacetimes described by GDG. This strongly suggests that the microscopic degrees of freedom that account for black hole entropy are Goldstone boson like excitations associated with a symmetry breaking of diffeomorphism invariance by the horizon boundary conditions.

The purpose of the present review is twofold. Firstly, we wish to review the Hamiltonian framework for GDG, with emphasis on the definition and properties of the generalized Misner–Sharpe mass [7] which plays an important role in the study of dynamical horizons [8]. Secondly, we will illustrate with a specific example [9] the utility of GDG in studying the quantum properties of black holes.

The chapter is organized as follows. In the next section we summarize the Lagrangian and Hamiltonian descriptions of GDG. Section III contains the definition of the generalized Misner–Sharpe mass function and reviews its fundamental properties. We then implement a partial gauge fixing that results in a local Hamiltonian constraint in which spatial points decouple. Section IV reviews the calculation of the quantum structure of black hole spacetime obtained by quantizing this Hamiltonian using a technique called polymer quantization that is motivated by loop quantum gravity. Section V closes with a summary and conclusions.

2. Lagrangian Formulation of GDG

Generic 2-D dilaton gravity (GDG) [2, 4] starts from the most general diffeomorphism invariant action consisting of a scalar (the dilaton) and metric that contains at most second derivatives of the fields:

$$S = S_G[g, \phi] + S_M, \tag{1}$$

where

$$S_G[g,\phi] = \frac{1}{2G}\int dx\,dt\,\sqrt{-g}\left(\phi R(g) + \frac{V(\phi)}{l^2}\right) \tag{2}$$

and l is a parameter with dimensions of length which is generally taken to be the Planck length in the theory. The matter action is arbitrary for much of what follows, but to be specific we consider the action for a massless scalar field non-minimally coupled to the dilaton:

$$S_M[\psi, g, \phi] = -\frac{1}{2}\int d^2x\sqrt{-g}\,h(\phi)|\nabla\psi|^2. \tag{3}$$

The dilaton is necessary in (2) because in two spacetime dimensions the Ricci scalar is a total divergence and the Einstein tensor vanishes identically. Up to conformal reparametrizations of the metric [3] the entire class of vacuum theories is thus parametrized by a single function in the action, namely the dilaton potential.

Crucial for any physically relevant discussion of quantum black holes is the fact this class of models contains spherically symmetric Einstein gravity [1] in arbitrary spacetime dimensions. One can verify that the action:

$$I^{(D)} = \frac{1}{16\pi G^{(D)}}\int d^Dx\sqrt{-g^{(D)}}\,R^{(D)} - \frac{1}{2}\int d^Dx\sqrt{-g^{(D)}}\,|\nabla\overline{\psi}|^2 \tag{4}$$

reduces to (2) providing one identifies the physical D-metric by:

$$ds^2_{phys} = \frac{ds^2}{j(\phi)} + r^2(\phi)d\Omega^2_n, \tag{5}$$

where $d\Omega^2_n$ is the line element of the unit n-sphere and

$$j(\phi) := \int_0^{\phi} d\widetilde{\phi}V(\widetilde{\phi}). \tag{6}$$

In additon one needs to make the following substitutions:

$$
\begin{aligned}
2G &= \frac{16\pi G^{(n+2)} n}{8(n-1)\nu^{(n)} l^n}, &(7)\\
\phi &= \frac{n}{8(n-1)}\left(\frac{r}{l}\right)^n, &(8)\\
V(\phi) &= (n-1)\left(\frac{n}{8(n-1)}\right)^{1/n} \phi^{-1/n}, &(9)\\
h(\phi) &= \frac{8(n-1)}{n}\phi = \left(\frac{r}{l}\right)^n, &(10)\\
\psi &= \sqrt{\nu^{(n)} l^n}\,\overline{\psi}. &(11)
\end{aligned}
$$

In keeping with our expectations from spherically symmetric gravity, GDG satisfies a generalized Birkhoff theorem [10]: the most general vacuum solution has a Killing vector, contains no local degrees of freedom contains a single free parameter. It can be written in Schwarzschild like coordinates $\widetilde{x} = l\phi$ as:

$$
ds^2 = -(j(\phi) - 2GMl)dt_s^2 + (j(\phi) - 2GMl)^{-1} d\widetilde{x}^2, \qquad (12)
$$

where M is the ADM mass. GDG can also easily accomodate an electromagnetic field and cosmological constant in the higher dimensional spacetime simply by suitably modifying the dilaton potential [11].

The correspondence in Eq. (11) is only valid for $D \geq 4$. By choosing the dilaton potential to be linear in ϕ, one obtains the Jackiw–Teitelboim theory [12]. This theory is equivalent to dimensionally reduced $D = 3$ Einstein gravity with cosmological constant which in turn contains the famous BTZ black hole [13]. Three dimensional gravity and the BTZ black hole have provided many useful insights into quantum gravity in general and quantum black holes [14]. Finally, we mention that a constant dilaton potential yields the string inspired CGHS model [15]. Interestingly, this model can also be obtained formally as the $n \to \infty$ of n-dimensional spherically symmetric gravity. (cf. (11).

3. Hamiltonian Analysis

Following the formalism in [4], we start with the metric as parametrized in modified ADM form:

$$ds^2 = e^{2\rho}\left(-\sigma^2 dt^2 + (dx + N dt)^2\right) \tag{1}$$

and assume that the $2 - D$ spacetime manifold $M_2 = R \times \Sigma$, where Σ is a spacelike slice. The standard analysis [4] reveals a Lagrangian of the form:

$$L(t) = \int dx \left(\Pi_\rho \dot{\rho} + \Pi_\phi \dot{\phi} + \Pi_\psi \dot{\psi}\right) - H(t) \tag{2}$$

with Hamiltonian $H(t)$ given by:

$$H(t) = (\sigma \mathcal{G} + N\mathcal{F}) + H_B. \tag{3}$$

In the above $\mathcal{G}$ is the Hamiltonian constraint,

$$\mathcal{G} = \frac{\phi''}{G} - \frac{\phi'\rho'}{G} - G\Pi_\phi\Pi_\rho - \frac{e^{2\rho}}{2G}\frac{V(\phi)}{l^2} + \mathcal{G}_M, \tag{4}$$

where $\mathcal{G}_M$ is the matter part of the Hamiltonian constraint:

$$\mathcal{G}_M = \frac{1}{2}\left(\frac{\Pi_\psi^2}{h(\phi)} + h(\phi)(\psi')^2\right). \tag{5}$$

The diffeomorphism constraint $\mathcal{F}$ also consists of a gravitational and matter part $\mathcal{F}_M$:

$$\mathcal{F} = \rho'\Pi_\rho - \Pi_\rho' + \phi'\Pi_\phi + \mathcal{F}_M \tag{6}$$

with $\mathcal{F}_M = \psi'\Pi_\psi$. For a given set of boundary conditions, the boundary term H_B is chosen in order to make the variational principle well defined.

As usual the lapse and shift functions σ and N, respectively, are lagrange multipliers that give rise to the Hamiltonian and diffeomorphism constraints in the theory. For future reference we write down the time derivatives of the fields that follow from (2):

$$\dot{\rho} = N\rho' + N' - \sigma G\Pi_\phi, \tag{7}$$

$$\dot{\phi} = N\phi' - \sigma G\Pi_\rho, \tag{8}$$

$$\dot{\psi} = N\psi' + \frac{\sigma\Pi_\psi}{h(\phi)}, \tag{9}$$

where the prime denotes partial differentiation with respect to the spatial coordinate x.

4. Generalized Misner–Sharpe Mass

A very useful form of the Hamiltonian constraint is obtained by taking the following linear combination of constraints:

$$\begin{aligned} \widetilde{\mathcal{G}} &:= l e^{-2\rho}\phi'\mathcal{G} + lGe^{-2\rho}\Pi_\rho\mathcal{F} \\ &= -\mathcal{M}' + \rho_M(\psi, \Pi_\psi, \rho, \Pi_\rho, \phi), \end{aligned} \tag{1}$$

where we have defined

$$\rho_M \equiv l\frac{e^{-2\rho}\phi'}{2}\left(\frac{\Pi_\psi^2}{h(\phi)} + h(\phi)(\psi')^2 + \frac{2G\Pi_\rho\Pi_\psi\psi'}{\phi'}\right). \tag{2}$$

As the notation suggests ρ_M is the energy density of matter, including gravitational self-interactions, of the spacetime. In the above

$$\mathcal{M} = \frac{l}{2G}\left(e^{-2\rho}\left((G\Pi_\rho)^2 - (\phi')^2\right) + \frac{j(\phi)}{l^2}\right) \tag{3}$$

is the generalized Misner–Sharpe mass function [7] for the class of theories considered here. The integrated form of (1) reveals that the Misner–Sharpe mass is equal to the total energy density, including gravitational self interaction, contained within a given spatial volume. The generalized Misner–Sharpe mass is therefore useful because of its natural interpretation as the quasi-local mass within a sphere of radius r. For boundary conditions corresponding to asymptotically flat spacetimes, $\mathcal{M}$ approaches the ADM mass at infinity. At null infinity it reduces to the Bondi-Sachs mass.

By using Eqs. (9) one can verify the following covariant expression for $\mathcal{M}$ [4]:

$$j(\phi) - 2G\mathcal{M} = -|\nabla\phi|^2. \tag{4}$$

This generalizes the 4-D definition of the Misner–Sharpe mass to generic 2-d dilaton gravity. In addition, the RHS of (4) can be written as the product of ingoing and outgoing null expansion [8,9,16]:

$$-|\nabla\phi|^2 = e^{-2\rho}\left((G\Pi_\rho)^2 - (\phi')^2\right) = \Theta^+\Theta^-, \tag{5}$$

where

$$\Theta^{(\pm)} = \frac{\nabla^\mu \phi}{\phi} l^{(\pm)}_\mu, \tag{6}$$

where the vectors

$$l^{(\pm)}_\mu = n_\mu \pm s_\mu, \tag{7}$$

are ingoing and outgoing null vectors defined in terms of the future pointing normal $n_\mu = e^\rho[\sigma, 0]$ of each spacelike slice defined by the metric (1), and the embedded inward pointing normal $s_\mu = e^\rho[-N, -1]$. A straightforward calculation yields:

$$\begin{aligned}\Theta^{(\pm)} &= \frac{e^{-\rho}}{\phi}\left(\frac{\dot{\phi}}{\sigma} + \phi'\left(\frac{N}{\sigma} \pm 1\right)\right) \\ &= \frac{e^{-\rho}}{\phi}(\Pi_\rho \pm \phi').\end{aligned} \tag{8}$$

from which we can recover (5).

Eqs. (4) and (5) verify that as in the 4-d case [8], the invariant $\mathcal{M}$ plays a vital role in studying dynamical horizons: it determines whether a surface is trapped ($j - 2G\mathcal{M} < 0$), untrapped ($j(\phi) - 2G\mathcal{M} > 0$) or marginal ($j(\phi) - 2G\mathcal{M} = 0$).

5. Partial Gauge Fixing

We now partially fix the gauge as in [9, 17] by choosing the spatial coordinate to be a specific function of the dilaton ϕ.

$$\phi' = j(\phi) \tag{1}$$

or equivalently:

$$x = \int_0^\phi \frac{d\widetilde{\phi}}{j(\widetilde{\phi})}. \tag{2}$$

For spherically symmetric gravity, this implies that $x = r$, the areal radius. This slicing does not in general cover the entire maximally extended spacetime of an eternal single horizon black hole. It can, however, be chosen to cover either the causal future of a past horizon, or the causal past of a future horizon. The former is of course relevant to the study of black hole formation via gravitational collapse [18, 19]. As well, (1) can be generalized to $\phi' = f(\phi)$ for arbitrary monotonic $f(\phi)$, but this complicates the discussion without resulting in any fundamental differences. Finally, we note also that our partial gauge fixing procedure is different from the "flat slice" choice of Husain and Winkler [20] who fix the conformal mode of the metric instead of ϕ.

From the equation for $\dot{\phi}$ in (9) one obtains the consistency condition for this gauge choice:

$$N = \frac{\sigma G \Pi_\rho}{\phi'} . \tag{3}$$

One can solve the diffeomorphism constraint to determine Π_ϕ so that the metric can be written in terms of two arbitrary functions, namely the ρ and the lapse σ. The result for D dimensional gravity is:

$$\begin{aligned} & ds^2_{phys} \\ &= \frac{e^{2\rho}}{j(\phi)}\left(-\sigma^2 dt^2 + \left(dx \pm \sqrt{1 - \frac{e^{2\rho}}{j(\phi)^2}\left(j(\phi) - 2G\mathcal{M}/l\right)}\,\sigma dt\right)^2\right) \\ &\quad + r^2(\phi) d\Omega. \end{aligned} \tag{4}$$

The mass function, $\mathcal{M}$, or equivalently the momentum Π_ρ conjugate to ρ, is determined implicitly in terms of ρ and the matter fields along a given spatial slice by solving the Hamiltonian constraint (1).

At this stage one can go to the completely reduced theory by specifying the conformal mode of the metric $e^{2\rho}$. There is a unique choice that eliminates the off-diagonal term in the metric:

$$e^{2\rho} = \frac{j(\phi)^2}{j(\phi) - 2G\mathcal{M}} . \tag{5}$$

This yields generalized Schwarzschild coordinates that are singular at horizons where $j(\phi) - 2G\mathcal{M} = 0$. All other choices produce slicings that are regular across either future or past horizons, depending on the choice of sign in front of the square root. In particular, one obtains the 2-d dilaton gravity anologue of flat slice or Painleve–Gullstrand coordinates by choosing $e^{2\rho} = j(\phi)$ [18, 19]. The metric then takes the form:

$$ds^2_{phys} = \left(-\sigma^2 dt^2 + \left(dx \pm \sqrt{\frac{2G\mathcal{M}l}{j}}\,\sigma dt\right)^2\right) + r^2 d\Omega, \qquad (6)$$

where the positive (negative) sign yields slicings that are regular across future (past) horizons. The lapse function σ is now also determined via a consistency condition for this gauge choice. Thus, as expected from the fact that dilaton gravity contains no propagating degrees of freedom, the metric is completely determined by the matter distribution, up to a constant of integration, which measures the strength of the singularity, if any, at the origin and contributes to the ADM mass at infinity. Note that $t = const$ surfaces in the physical higher dimensional metric (5) are flat.

6. Polymer Quantization of the Vacuum Theory

We now restrict consideration to the vacuum theory for which the Hamiltonian constraint reduces to:

$$\mathcal{M} = M, \qquad (1)$$

where M spatially constant and a constant of motion. This is consistent with the fact that the completely reduced vacuum theory contains two phase space variables: the mass M and its conjugate momentum P_M [21, 22] which has a geometrical interpretation as the Schwarzschild time separation between opposite ends of the spatial slices.

An important feature of this partial gauge fixing is that the mass function does not contain spatial derivatives of the remaining degrees of freedom. By performing the canonical transformation:

$$X = e^{\rho}, \qquad (2)$$

$$P = e^{-\rho}\Pi_\rho, \qquad (3)$$

we can express the mass function in the form:

$$\mathcal{M} = \frac{l}{2G}\left[G^2P^2 - \frac{j(\phi)^2}{X^2} + \frac{j(\phi)}{l^2}\right]. \qquad (4)$$

In the following, we present the results of the so-called "Dirac quantization" procedure in which the Hamiltonian constraint is imposed on physical states as an operator constraint. Details can be found in [9]. The operator constraint equation that needs to be solved is:

$$\left(G^2\hat{P}^2 - \frac{j^2}{\hat{X}^2}\right)|\Psi\rangle = \epsilon(\phi, M)|\Psi\rangle. \tag{5}$$

Remarkably, at each spatial point (5) describes the Hamiltonian of a particle moving in a $1/X^2$ potential, with coupling $j(\phi)^2$ and energy $\epsilon(\phi, M) \equiv (2GM - j(\phi)$. The $1/X^2$ potential arises in a variety of contexts, including the quantum Hall effect [23] and a field theory formulation of nucleon interactions [24]. An important property of the inverse square potential Hamiltonian is its scale invariance. As a result, it possesses three symmetry generators that span an algebra isomorphic to the conformal group $SO(1,2)$. The quantum mechanics of this potential has been the subject of much work [25], in part because the conformal symmetry is necessarily broken at the quantum level.

Note that the eigenvalue ϵ in (5) is negative in the asymptotic region outside the horizon ($j(\phi) > 2GlM_0$, so that one is looking for bound states in this region. In the interior one is looking for "scattering states".

We now quantize the partially gauge-fixed theory in the discrete Bohr quantization framework following [16,27]. The Bohr quantization of the $1/X^2$ quantum mechanical potential was also considered more recently in [26]. We first define a basis $|a_1, a_2, \ldots, a_N\rangle$ of orthonormal states, such that:

$$\left\langle a'_1, a'_2, \ldots, a'_N | a_1, a_2, \ldots, a_N \right\rangle = \delta_{a'_1,a_1}\delta_{a'_2,a'_2}\cdots, \tag{6}$$

on which we define the action of the quantum smeared operator $\hat{X}_f$:

$$\hat{X}_f|a_1, a_2, \ldots, a_N\rangle = L\sum_i (f(x_i)a_i)|a_1, a_2, \ldots, a_N\rangle, \tag{7}$$

where:

$$X_f \equiv \int_\Sigma dx\, f(x)X(x) \tag{8}$$

for some suitable test function $f(x)$. L is a constant of dimension length which will be determined below. The discretization $\{x_i\}$ is at this stage completely arbitrary. The a_i are the eigenvalues of $\hat{X}$ at the point x_i. This spectrum is by construction discrete, and will be determined once we choose a representation for its conjugate momentum, $\hat{P}$.

The operator $\hat{P}$ can't be defined as a self-adjoint operator because the discrete topology assigned to the eigenstates of $\hat{X}$ prevents the implementation of infinitesmal transformations generated by $\hat{P}$. It is however possible to define the discrete translation operator: $\hat{U}_\gamma(x)$ whose commutator with $\hat{X}$ is:

$$[X_f, U_\gamma(x)] = i\hbar\gamma G l f(x) U_\gamma(x). \tag{9}$$

where γ is dimensionless. The action of $\hat{U}_\gamma(x)$ on the basis states is:

$$\hat{U}_\gamma(x_i)|a_1, a_2, \ldots, a_N\rangle = |a_1, \ldots, a_i - \gamma, \ldots, a_N\rangle. \tag{10}$$

The definitions (7) and (10) imply the quantum commutator:

$$[\hat{X}_f, \hat{U}_\gamma(x_i)] = -\gamma L \hat{U}_\gamma(x_i). \tag{11}$$

A comparison of (9) and (11) determines the arbitrary constant L to be proportional to the Planck scale:

$$L = \hbar G l \tag{12}$$

It is now possible to determine the spectrum of $\hat{X}$ as required by the action of $\hat{U}_\gamma(x_i)$:

$$\begin{aligned} |a_1, \ldots, a_i, \ldots, a_N\rangle &= (\hat{U}_{\pm\gamma}(x_i))^m |a_1, \ldots, a_i^{(0)}, \ldots, a_N\rangle \\ &= |a_1, \ldots, a_i^{(0)} \pm m\gamma, \ldots, a_N\rangle. \end{aligned} \tag{13}$$

Given the physical interpretation of X in terms of the square root of the conformal mode of the metric, it is reasonable to assume that the spectrum is symmetric about 0, and take $a_i^{(0)} = 0$. Henceforth we assume that $a_i = m\gamma$ where m is any integer.

In order to implement the Hamiltonian constraint we define $\hat{P}^2(x_i)$ in terms of the discrete translation operator [27]:

$$\begin{aligned} l^2 \hat{P}^2(x_i)|a_1, \ldots, a_N\rangle &\equiv \lim_{\gamma\to 0} \frac{1}{\gamma^2} \left(2 - U_\gamma(x_i) - U_{-\gamma}(x_i)\right) |a_1, \ldots, a_N\rangle \\ &= \lim_{\gamma\to 0} \frac{1}{\gamma^2} \Big(2|a_1, \ldots, a_N\rangle - |a_1, \ldots, a_i - \gamma, \ldots, a_N\rangle \\ &\qquad - |a_1, \ldots, a_i + \gamma, a_N\rangle \Big). \end{aligned} \tag{14}$$

Finally, we can implement the smeared Hamiltonian constraint on the basis states:

$$\hat{O}_f|a_1,\ldots,a_N\rangle = \frac{1}{2}\int dx f(x)\left(l^2G^2\hat{P}^2 - \frac{j(\phi)^2}{\hat{X}^2} - \epsilon(\phi)\right)|a_1,\ldots,a_N\rangle$$
$$\equiv \frac{L}{2}\sum_i f(x_i)\left(G^2l^2\hat{P}^2 - \frac{j_i}{a_i^2} - \epsilon_i\right)|a_1,\ldots,a_N\rangle. \quad (15)$$

where for ease of notation we henceforth denote $j_i := j(\phi(x_i))$ and similarly for ϵ_i.

We look for physical states: $|\psi\rangle$:

$$|\psi\rangle = \sum_{a_1,\ldots,a_n} \psi(a_1,\ldots,a_N)|a_1,\ldots,a_N\rangle. \quad (16)$$

that solve the Hamiltonian constraint

$$\hat{O}_f|\psi\rangle = 0, \quad (17)$$

where Given the absence of coupling between spatial points in the Hamiltonian constraint, it is possible to separate variables:

$$\psi(a_1,\ldots,a_N) = \psi_1(a_1)\psi_2(a_2)\cdots\psi_N(a_N). \quad (18)$$

After some algebra one obtains the following difference equation at each spatial point:

$$\left[\psi^{m+1} + \psi^{m-1} - \left(2 - \epsilon\gamma^2\right)\psi^m\right] + \frac{\gamma^2 j^2}{(m\gamma)^2}\psi^m = 0. \quad (19)$$

where we have suppressed the index i labelling the spatial point and used the fact that $a_i = m\gamma$ to write $\psi(a) = \psi^m$ so that $\psi(a\pm\gamma) = \psi^{m\pm1}$. Note that while the potential in the last term on the LHS of (19) is singular at $m = 0$. This singularity can be handled either by regulating the potential using the analogue of the Thiemann trick [28] or by imposing suitable boundary conditions on the states [26] with qualitatively similar results. We choose the latter and set $\psi^0 = 0$, which corresponds to the anti-symmetric boundary conditions of [26].

A discrete spectrum is obtained by requiring the physical states to be normalizable:

$$\langle\psi\rangle^2 := \sum_m |\psi^m|^2 < \infty. \tag{20}$$

The difference equation (19) can be rewritten as

$$\begin{aligned} \psi^0 &= 0, && (21)\\ \psi^{n+1} + \psi^{n-1} &= \left(2a - \frac{b}{n^2}\right)\psi^n, \quad n = 1, 2, \ldots && (22) \end{aligned}$$

with

$$\begin{aligned} a &:= 1 + (j(\phi) - 2GM\ell)\gamma^2/2; \\ b &:= j^2(\phi). \end{aligned} \tag{23}$$

This equation can solved for the eigenvalues and eigenstates either by using "shooting method" as in reference [26] or by establishing an asymptotic expansion for each of the infinitely many allowed values of the spatial coordinate as parametrized by the quantity $j(\phi)$. In the latter case one obtains an infinite number of self-consistency conditions, fixing the spectrum. We refer the reader to [9] for details. The results are contained in Figure 1 (1), which we now summarize.

The solid lines plot the allowed values of a and b obtained from the consistency conditions implied by the normalizability condition on the wave function.Note that $a = 1$ corresponds to the location of the horizon. It is clear that for each fixed value of b, which is the coupling constant in the Hamiltonian, there is an infinite tower of possible states. In this Figure, the ground state for each value of b occurs at the top of the figure, whereas the excited states "pile" up near $a = 0$, which corresponds to the zero energy (semi-classical) end of the bound state spectrum. The spectrum of allowed a for fixed coupling b agrees with the results in [26]. However, in the present case, a and b are not independent, since for fixed black hole mass M_0 and discretization parameter γ they are both determined by the value of the dilaton ϕ. In particular, by eliminating $j(\phi)$ we see that Eqs.(23) imply that:

$$b = \left(\frac{2(a-1)}{\gamma^2} + 2GMl\right)^2. \tag{24}$$

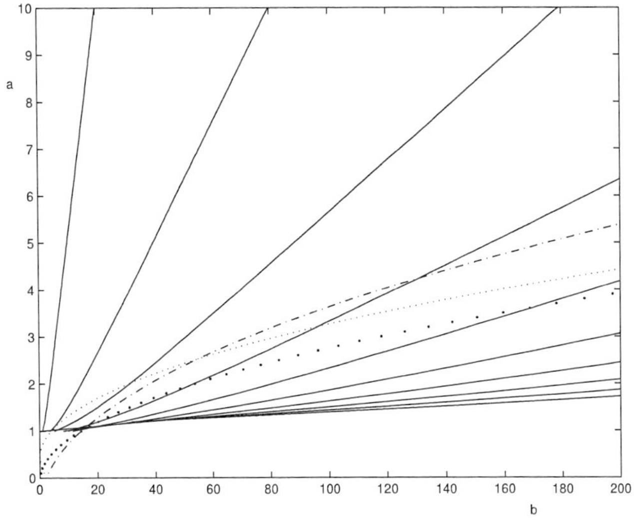

Figure 1. Solid lines graph the relations between a and b from asymptotic expansions in Table 1. The parabolas graph allowed r vs b for various M and $1/\gamma^2$. The large dotted line is $2GMl = 1.90$, $1/\gamma^2 = 1.77$; the small dotted line is $2GMl = 1.00$, $1/\gamma^2 = 1.77$; and finally the dot-dash line is $2GMl = 1.90$, $1/\gamma^2 = 1.18$. The horizon is at $a = 1$. Intersection points of parabolas with the solid lines yield the spectrum for $j(\phi)$ either in terms of $a := 1 + \frac{1}{\gamma^2}(j(\phi) - 2GM\ell)$ or of $b = (j(\phi))^2$.

The parabolae plotted with dotted lines give representative relations between a and b that come from there physical interpretation. Thus, the allowed values for $j(\phi)$ and hence ϕ and implicitly r are given by the location of the intersection points between a particular parabola (for fixed M_0 and γ) and the solid lines that provide the consistency conditions for normalizable states.

Figure 1 shows qualitatively the effects of changing the mass M of the black hole and the discretization scale γ. Increasing the mass shifts the parabola down (small to large dots) and the spectrum for a is squeezed closer to the horizon ($a = 1$). Decreasing γ (dot-dash to large dots) flattens the parabola and moves the eigenvalues of $b = (j(\phi))^2$ closer together. This behaviour corresponds to our physical expectation that the spatial topology near the horizon becomes more continuous with increasing mass or decreasing γ.

Note that the mass is expressed in terms of Planck units. For very large M, one expects the spectrum in the exterior of the black hole to be approximately continuous even near the horizon. This in fact can be seen to happen by noting from that as MGl increases, the horizon $a = 1$ occurs for larger and larger b, which in turn implies that the parabola first intersects the asymptotic expressions in Figure 1 for larger and larger m, where the lines are more and more dense. Thus the lowest eigenvalues get closer and closer together as MGl increases.

In addition, the consistency condition curves that for large a and b straighten out and their slopes happen to be $2m^2$ for integers $m = 1, 2, \ldots$. This implies that far from the horizon

$$j = m^2\gamma^2. \tag{25}$$

Using (25) and the fact that $j \gg 2GMl$, we find:

$$\frac{\Delta j}{j} = \frac{2}{m} \ll 2\frac{\gamma}{\sqrt{2GMl}}, \tag{26}$$

which is small for macroscopic black holes so that the spatial spectrum approaches the continuum asymptotically as expected.

7. Conclusion

Spherically symmetric gravity and its natural generalization 2-D generic dilaton gravity provide an ideal theoretical laboratory for the study of quantum corrections to the properties of black hole spacetimes. We have presented just a couple of results that have emerged from a study of the vacuum theory. The first, and perhaps most significant, was the work of Carlip [5,6] suggesting that the microscopic source of the Bekenstein–Hawking entropy might be found in spherically symmetric diffeos that become physical when horizon boundary are imposed. This is encouraging because it indicates that even though imposing symmetry removes the gravitons from the theory, we have not thrown away the "baby with the bath water". It also bodes well for other models without gravitons, such as 2+1 dimensional gravity, that are currently studied in this context.

We also showed how the Misner–Sharpe mass, originally defined for spherically symmetric gravity, is easily generalized to GDG and expressed either covariantly or in terms of the phase space variables.

Finally, we presented the results of the polymer quantization scheme applied to the the spacetime surrounding a generic, single horizon, spherically symmetric black hole. The key features of the scheme are:

- Only the spatial diffeomorphisms were fixed at the classical level.
- The single remaining constraint (the Hamiltonian constraint) was imposed as an operator constraint on physical states in the quantum theory.
- In the class of partial gauge fixings considered the constraint did not contain At each point the quantum mechanics was that of a particle moving in a $\frac{1}{X^2}$ potential, with the analogue of particle position being the $X(x)$ square root of the conformal mode of the metric.
- The condition that the amplitude at $X(x) = 0$ to vanish tamed the singularity in the potential.
- The requirement that physical states (i.e. those annihilated by the constraint) be normalizable gave rise to a discrete spatial spectrum exterior to the horizon with reasonable semi-classical properties: it approximates the continuum far from the horizon even for microscopic black holes, as well as near the horizon for macroscopic black holes.
- One can speculate on the significance of the continuum of states near the horizon. The Hamiltonian at each spatial point is scale invariant and there exists a set of $SO(2,1)$ symmetry generators that are defined classically in terms of the operators X and P. This symmetry is necessarily broken at the quantum level. It may be that this symmetry breaking near the horizon is connected to the statistical mechanical entropy of the black hole, much as in the work of [5, 6].

There are many other interesting results that have emerged from vacuum GDG and also GDG with matter that we have not been able to discuss due to lack of space. Some of these are described in [29]. We hope however to have convinced the reader that theses theories are relevant to the study of black hole quantum mechanics and worthy of further investigation.

Acknowledgments

G.K. thanks the Perimeter Institute for its kind hospitality during the initial stages of the preparation of this chapter. This work was supported in part by the Natural Sciences and Engineering Research Council of Canada.

References

[1] B.K. Berger, D. M. Chitre, Y. Nutku, V.E. Moncrief, *Phys. Rev. D* **5** 2467–2470 (1972); W. Unruh, *Phys. Rev. D* 14, 870 (1976).

[2] D. Grumiller, W. Kummer, D.V. Vassilevich, *Phys. Rept.* 369 327–430 (2002).

[3] T. Banks and M. O'Loughlin, *Nucl. Phys. B* 362 649 (1991).

[4] D. Louis-Martinez, J. Gegenberg and G. Kunstatter, *Phys. Lett. B* 321 193 (1994), arXiv:gr-qc/9309018; J. Gegenberg, G. Kunstatter and D. Louis-Martinez, *Phys. Rev. D* 51 1781–1786 (1995).

[5] S. Carlip, *Phys. Rev. Lett.* 88 241301 (2002).

[6] S. Carlip, *Phys. Rev. Lett.* 99 021301 (2007).

[7] C.W. Misner and D.H. Sharpe, *Phys. Rev.* 136 (1964).

[8] S.A. Hayward, *Phys. Rev. D* 53 1938 (1996); *Phys. Rev. D* 49 6467 (1994); *Phys. Rev. D* 70 104027 (2004).

[9] J. Gegenberg, G. Kunstatter and D. Small, *Class.Quant.Grav.* 23 6087–6100 (2006), arXiv:gr-qc/0606002.

[10] D. Louis-Martinez and G. Kunstatter, *Phys. Rev. D* 49 5227–5230 (1994).

[11] A.J. Medved and G. Kunstatter, *Phys. Rev. D* 59 104005 (1998).

[12] R. Jackiw, in *Quantum Theory of Gravity*, edited by S. Christensen (Hilger, Bristol, 1984), p. 403; C. Teitelboim, *ibid* p. 327.

[13] M. Bañados, C. Teitelboim, J. Zanelli, *Phys. Rev. Lett.* 69 1849–1851 (1992). hep-th/9204099.

[14] S. Carlip, "Quantum Gravity in $2+1$ Dimensions", Cambridge University Press (2003).

[15] C.G. Callan, S.B. Giddings, J.A. Harvey and A. Strominger, Phys. Rev. D 45 R1005 (1992). See also S. Hawking, *Phys. Rev. Lett.* 69 406 (1992).

[16] V. Husain and O. Winkler, *Phys. Rev. D* 69 084016 (2004), arXiv:gr-qc/0312094; V. Husain and O. Winkler, *Class. Quant. Grav.* 22 L127–L134 (2005), arXiv:gr-qc/0410125.

[17] R. Daghigh, J. Gegenberg and G. Kunstatter, *Class. Quant. Grav.* 24 2099–2107 (2007). gr-1c/0607122.

[18] J. Ziprick and G. Kunstatter, *Phys. Rev.* D80 024032 (2009). arXiv:0902.3224 [gr-qc].

[19] J. Ziprick and G. Kunstatter, *Phys. Rev. D* 79 101503(R) (2009), arXiv:0812.0993[gr-qc] .

[20] V. Husain and O. Winkler, *Phys. Rev. D* 71 104001 (2005), arXiv:gr-qc/0503031; V. Husain and O. Winkler, *Phys.Rev.* D73 124007 (2006) arXiv:gr-qc/0601082.

[21] K. Kuchar, *Phys. Rev. D* 50 3961–3981 (1994).

[22] D. Louis-Martinez, J. Gegenberg and G. Kunstatter, *Phys. Lett. B* 321 193–198 (1994). gr-qc/9309018.

[23] R. M. Cavalcanti and C.A.A. de Carvalho, *J. Phys. A Math. Gen.* 31 2391 (1998); 32 (1999).

[24] S. Weinberg, Physica A 96 327 (1990); *Nucl. Phys. B.* 363 (1991); C. Ordonez, L. Ray and U. van Koclck *Phys. Rev. C P* 53 2086 (1996).

[25] H. E. Camblong, C. R. Ordonez, *Phys. Rev. D* 68 125013 (2003); H. E. Camblong, L.N. Epele, H. Fanchiotti and C.A. Garcia *Canal Annals Phys.* 287 14–56 (2001); *Phys. Rev. Lett.* 85 1590–1593 (2000).

[26] G. Kunstatter, J. Louko and J. Ziprick, *Phys. Rev. A* 79 032104 (2009); arXiv:0809.5098[gr-qc].

[27] A. Ashtekar, S. Fairhurst and J.L. Willis, *Class. Quant. Grav.* 20 1031–1062 (2003), arXiv:gr-qc/0207106.

[28] T. Thiemann, *Class. Quant. Grav.* 15 839 (1998), arXiv:gr-qc/9606089.

[29] J. Gegenberg and G. Kunstatter, "2-D Midisuperspace Models for Quantum Black Holes", in 'Fundamental Interactions: A Memorial Volume for Wolfgang Kummer,' Editors: Daniel Grumiller, Anton Rebhan and Dimitri Vassilevich, World Scientific, 2010, pp.231-247; arXiv:0902.0292*Wolfgang Kummer Memorial Volume*, (World Scientific, 2009) eds. D. Grumiller, A. Rebhan and D. Vassilevich.

In: Recent Research in Quantum Gravity ISBN: 978-1-61942-386-2
Editor: A. Dasgupta, pp. 115-135

Chapter 5

IS THE SPACETIME METRIC EUCLIDEAN RATHER THAN LORENTZIAN?

Rafael D. Sorkin*
Perimeter Institute, Waterloo ON, Canada
Department of Physics, Syracuse University
Syracuse, New York, US

Abstract

My answer to the question in the title is "No". In support of this point of view, we analyze some examples of saddle-point methods, especially as applied to quantum "tunneling" in nonrelativistic particle mechanics and in cosmology. Along the way we explore some of the interrelationships among different ways of thinking about path-integrals and saddle-point approximations to them.

1. Introduction

The use of non-Lorentzian metrics (complex metrics or metrics of Euclidean signature) as instantons or in connection with black hole thermodynamics has sometimes provoked the opinion that spacetime is "really" a Riemannian geometry as opposed to a Lorentzian one; and given this, it may not be out of place to

*E-mail address: sorkin@physics.syr.edu

adduce here certain more or less obvious points to the contrary. In the following I will illustrate these points with some simple and — I hope — instructive examples of saddle-point approximations of the type that arise in "tunneling" processes, including the process of the birth of a cosmos, as it has been studied in "quantum cosmology". Much of what I will say is well known, but perhaps some of it is not as familiar as it might be, and I hope that the examples can help to clarify certain confusing aspects that arise in connection with saddle-point methods and the analytic continuation of path integrals. Perhaps also, the cosmology example can serve to bring out the conceptual (and potentially technical) advantage to be gained in quantum gravity by unambiguously taking all amplitudes to be defined in the first instance by Lorentzian path integrals.

Of course, it would be rash to reject dogmatically the possibility that quantum gravity might ultimately teach us to think of spacetime as more like a Riemannian geometry than a Lorentzian one, but it is fair, I think, to maintain that the usefulness of complex metrics as instantons provides no evidence on the question one way or the other. Rather, it seems much more straightforward and natural to interpret complex spacetimes simply as contributors to an analytically continued path integral that originally is taken over a space of Lorentzian metrics. If that interpretation is correct then for example, an amplitude computed from a Euclidean-signature instanton should be thought of as compounded from the contributions of a large number of Lorentzian histories, and the instanton itself should be thought of as merely a saddle-point of the analytically continued integrand.

2. The Saddle-Point Method as Applied to a Simple Integral

Below, I will illustrate what I mean, using the example of tunneling in nonrelativistic quantum mechanics, but to lay the groundwork for that discussion, it seems appropriate to recall in some detail how the saddle point method works in the one-dimensional case, as expounded in textbooks covering complex analysis [1]. This method is designed for integrals of the general form

$$\int g(x)\, dx \, \exp(f(x)) \,, \tag{1}$$

where the exponent $f(x)$ is supposed to be "rapidly varying" and $g(x)$ (the "measure-factor" in path-integral argot) is supposed to be "slowly varying with respect to $f(x)$". An instructive instance of this general form for us is

$$I(A) = \int_0^\infty \frac{dx}{x^{3/2}} \exp\left\{A(ix - 1/x)\right\}, \tag{2}$$

where A is an arbitrary positive parameter corresponding to $1/\hbar$ in the path-integral. We have then $f(x) = A(ix - 1/x)$ and $g(x) = x^{-3/2}$. In order to satisfy the condition of "rapid variation", I will assume further that $A \gg 1$. Not only does this integral bear a strong resemblance to a path-integral, but it can be evaluated exactly as

$$I(A) = \sqrt{\frac{\pi}{A}} \exp\left(\sqrt{2}\, A(i-1)\right), \tag{3}$$

thanks to our specific choice of $g(x)$. This, and not any specific analogy with a path-integral, was the reason why I chose $3/2$ as the power of x. Incidentally, the resemblance to a path-integral would have been even closer had we taken the exponent to be $iA(x - 1/x)$, but then (2) would have failed to be absolutely convergent.

The integrand in (2) analytically continues to the entire complex plane C except for $z = 0$, where it has an essential singularity. Its saddle points are by definition the solutions of $f'(z) = 0$, namely $z_\pm = \pm\sqrt{i} = \pm(1+i)/\sqrt{2}$, and at these points we have $f(z) = 2izA$, $f''(z) = 2Az$, $f'''(z) = -6A$. As a rule of thumb, a saddle-point approximation is valid when $|f'''(z)|^2 \ll |f''(z)|^3$ at the saddle-point, and this holds in consequence of our assumption that $A \gg 1$.

Since there are two saddle points, the question arises which if either we should use, but let's leave that aside for a moment. To each saddle point belongs a potential contribution to the integral, given up to sign by the general formula

$$g(z)\sqrt{\frac{2\pi}{-f''(z)}}\; e^{f(z)}, \tag{4}$$

or in our case,

$$\frac{1}{z^{3/2}}\sqrt{\frac{2\pi}{-f''(z)}}\; e^{f(z)}. \tag{5}$$

Remarkably, the saddle point at $z = +\sqrt{i}$ yields the exact answer. This accidental exactness would presumably not have happened had we chosen to absorb the prefactor $x^{-3/2}$ into the exponent $f(z)$. if we adopt the correct sign for the square root in (5). But how do we know which sign to choose without knowing the answer in advance, and how do we know which of the saddle points actually contributes? Both questions can be decided by ad hoc considerations in this case, but only if we already know that a saddle point approximation is actually valid. Concerning the question of which saddle point, one can notice first of all that $I(A) \to 0$ as $A \to \infty$, and compare this with the leading behavior of the two respective exponentials, namely $\exp(f(z\pm)) = \exp(\pm 2Ai^{3/2})$. Since $\exp(-2Ai^{3/2})$ blows up with A, only the saddle point at $z = z_+ = \sqrt{i}$ can be correct. The question of the sign can be decided by observing that $I(A)$ becomes a positive number as $A \to 0$, but this is even more of a cheat since $A \ll 1$ is altogether outside the domain of validity of the saddle point approximation. (Notice in this example that the correct saddle point is the one at which the integrand is smaller, not larger.)

Now let us approach the question more systematically. We started with an integral whose path or "contour" $\Gamma = \Gamma_0$ was the positive real axis, traversed from 0 to ∞. We then analytically continued the integrand and found two complex saddle points $z_\pm$. In order to make use of a saddle point, however, we need to deform Γ to pass through it, and it is necessary that this deformation not alter the value of the integral. In particular this means that Γ should not cross a singularity of the integrand (or if it does the extra contribution should be evaluated). In our case, this only has the effect of preventing us from rotating Γ freely about its initial endpoint, i.e. it requires the deformed contour to depart from the origin in the direction of positive x. To this extent, both saddle points are accessible.

However, the next condition is that, along the deformed contour Γ, the biggest contribution to the integral arise in the neighborhood of the saddle point, and not somewhere else. Normally, one tries to ensure this by arranging that away from the saddle point, either the integrand dies out rapidly (steepest descent method) or it suffers cancellation from rapid variations in its phase (stationary phase method). For the former method, the most favorable path through the saddle point is that of "steepest descent", along which the imaginary part of $f(z)$ remains constant. Along this path the phase of $\exp(f(z))$ remains constant, while its magnitude drops off rapidly as one departs from the saddle point. Given that $f''(z) \neq 0$ at the saddle point, there will be exactly one such path passing through it. In our case, and for the saddle point z_+, this path, a solution of the equation $\Im f(z) = constant$, or

$$-y - \frac{x}{x^2 + y^2} = constant = -\sqrt{2}\,, \tag{6}$$

originates at $z = 0$ with zero slope, curves upward passing through $z = z_+$, and goes off to infinity in the direction parallel to the positive imaginary axis. Clearly it is possible to deform such a path to our original contour Γ_0 by adjoining a large circular arc near infinity, and this won't affect our integral since our integrand drops off rapidly there. (See Figure 1) Hence, the saddle point $z = z_+$ is indeed "accessible", and we obtain the approximation (5) to $I(A)$. Finally, the unknown sign in (5) is determined by the rule that the phase of the square root factor should coincide with that of the tangent direction to the steepest descent contour Γ at the saddle point. The latter is easily evaluated and leads to the answer (3).

We could also have obtained this answer using a stationary-phase contour through $z = z_+$. Along such a curve, it is the real part of $f(z)$ that remains fixed, leading to an integrand of varying phase and (almost) constant magnitude. It would have been exactly constant if not for the prefactor $g(z)$ in the integrand. Through any saddle point (meaning a point where f' vanishes, but not f''), there will be two paths of constant $\Re f(z)$ meeting at right angles and rotated by 45 degrees from the steepest descent path. This makes a total of three rays in the tangent space. The fourth ray needed to complete the pattern (the one orthogonal to the steepest descent ray) might be called the direction of steepest *ascent*. It represents in some sense the worst possible choice for an integration contour. In our case, there is such a contour that rises vertically from $z = 0$, bends over to pass through z_+, and continues on to ∞, moving parallel to the x-axis. This also is a valid deformation of our original contour and leads to the same result (3). (For a stationary-phase path, one can determine the sign by applying the aforementioned rule to the "adjacent" steepest descent direction.)

Now what about the other saddle point z_-? For consistency it ought *not* to be accessible, since if it were, we would have two disagreeing approximations to the same integral. To see what goes wrong consider first the possible steepest descent contours through $z = z_-$. There is a curve of constant $\Im f(z)$ symmetric to the steepest descent curve through $z = z_+$, but it actually is a curve of "steepest ascent", as shows up clearly in Figure 1. The true steepest descent contour is the other curve of constant $\Im f(z)$ through $z = z_+$. It emerges from the origin horizontally to the right, bends downward and around to pass through $z = z_-$, and then continues upward to infinity parallel to the imaginary axis. By adding a circular arc at infinity as before, we could draw this curve down to the positive real axis, but the resulting contour would not be deformable to our original contour Γ_0, since it now would wind once around the origin. In order to unwind it, we would have to transport its "endpoint at infinity" in the opposite (counterclockwise) direction to what we just considered, but then the added circle at infinity would make an exponentially great contribution to the integral, as one sees in Figure 1. (Instead of adding an arc to the steepest descent contour near infinity, we could deviate from it at finite radii, but as Figure 1 shows, we would still have to pass through a region where the integrand would be exponentially greater than in the neighborhood of the saddle point.) We might also consider using one of the stationary-phase contours through $z = z_-$, but we would run into essentially the same obstruction with them as well. Either the contour would wind around the origin or the main contribution would fail to (be guaranteed to) come from the neighborhood of the saddle point.

I probably have lavished on this simple calculation more explanation than it deserves, but I wanted to illustrate that the mere existence of a saddle point is no guarantee of a corresponding saddle-point approximation. Rather, as we have seen, several conditions need to be satisfied which allow one to deform the original integration contour into one passing through the saddle point (or points) in such a manner as to validate the approximation (4). The cases of functional- or path-integrals are much more complicated, but it should be borne in mind that analogous conditions apply to them as well. As far as I know, these conditions are generally ignored entirely in the case of gravity, and people have never even tried to formulate them in that case.

In this connection, let us have a look at a different technique of analytic continuation which can also shed light on the quantum mechanical and gravitational cases, and which, when it applies, is in some ways simpler. In our integral (2), rather than analytically continuing the function f in its argument x, we can contemplate instead continuing in the parameter A. Actually, it's more convenient to rescale x or equivalently to introduce a second parameter that multiplies only

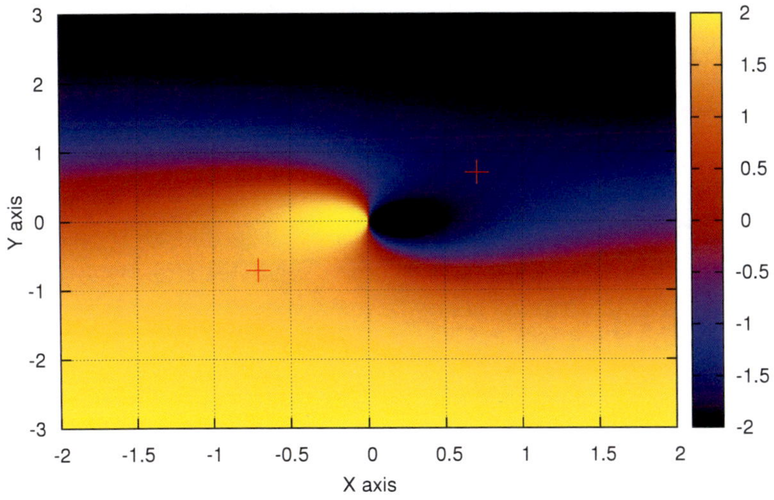

Figure 1. The colors depict the real part of the exponent $f(z)$, the range $(-\infty, \infty)$ having been compressed into $(-2, 2)$. The saddle points $z_{\pm}$ are marked with plus signs.

the first term in f. Our integral then takes the form $I(A) = J(B)B^{-1/4}$, where $B = A^2$ and

$$J(B) = \int_0^\infty \frac{dx}{x^{3/2}} \exp(iBx - 1/x) \,.$$

If now we analytically continue B to positive imaginary values, the integrand will become purely real and will acquire a saddle-point. In order that $J(B)$ be holomorphic in B the integral must continue to exist, however, and in order that the integral continue to exist, we need to continue B through the upper half-plane. This fixes the direction of the continuation. When $B = ib$ with b real and positive, the exponent, $-(bx + 1/x)$, has a single saddle point at $x = 1/\sqrt{b}$, and its value there is simply $-2\sqrt{b}$ or in other words $-2\sqrt{-iB}$.

(The point $x = -1/\sqrt{b}$ needn't be considered at all since it is outside the range of integration.) It is now easy to evaluate the contribution of this saddle-point to $J(B)$, and there is no longer any question of a sign ambiguity. The only small subtlety arises when we analytically continue the answer,

$$J(B) = \sqrt{\pi}\, \exp(-2\sqrt{-iB})\ ,$$

back from imaginary to real B (a well-defined procedure because we know $J(B)$ on the entire positive imaginary axis). Being careful to move in a clockwise direction so that B remains in the upper half-plane, we find that the exponent becomes $+\, i^{3/2} 2\sqrt{B} = \sqrt{2}\, A(i-1)$, and we recover our previous answer (3). Here, I have ignored a second subtlety, perhaps less small. How do we know that our approximate answer will still be approximately correct after we have continued it back to real B?

3. One Dimensional Tunneling

With the above example in mind, let us turn now to the case of one-dimensional tunneling through a potential barrier. Imagine, then, a (non-relativistic) particle emitted at $x = x_0$ by a "source of energy E" and absorbed by a "sink of energy E" at $x = x_1$. The amplitude for this process is formally

$$\int\limits_{\gamma\in\Gamma} d\gamma\, \exp\left\{\frac{i}{\hbar} S(\gamma)\right\}\ , \tag{7}$$

where

$$S(\gamma) = \int\limits_{\gamma} \frac{m}{2}\frac{dx^2}{dt} - V(x)dt\ +\ E(t_1 - t_0)\ . \tag{8}$$

Here t_0 and t_1 are respectively the times of emission and absorption, and the infinite-dimensional "integration contour" $\Gamma = \Gamma(x_0, x_1)$ comprises all trajectories γ from x_0 to x_1 (t_0 and t_1 being left free). When we come to analytically continue in the trajectory γ, it will be useful to have expressed $S(\gamma)$ also in so-called parameterized form. The distinction between the "parameterized" and "unparameterized" forms of the path-integral is not an independent question of it own, but an aspect of the question of "choice of measure-factor". i.e. as a worldline in R × R, coordinatized by an arbitrary parameter λ that we can take to run from 0 to 1. So expressed, S takes the form

$$S = \int_0^1 d\lambda \left(\frac{m\dot{x}^2}{2\dot{t}} + (E - V(x))\,\dot{t} \right) . \tag{9}$$

Now the integral (7) is not only formal, but even formally it is infinite because an infinite amplitude accumulates in an infinite time. In a more complete treatment, we would have to deal with this problem of normalization, but here my only interest is in the role played by the complexification of t. I will therefore ignore this issue together with all other technical issues that seem irrelevant to the main question (including for example the question of the "measure factor" hidden within $d\gamma$, the analog of $g(x)$ in our earlier example).

Now as usual, the choice of E divides R into regions that are "classically allowed" (where $E - V(x) > 0$) and "classically forbidden" (where $E - V(x) < 0$). Let us assume to begin with that both x_0 and x_1 are either within or on the edge of the forbidden region (which we assume to be connected). If they instead $[x_0, x_1]$ lay in the allowed region, then a classical trajectory γ_0 would join x_0 to x_1 and (such a trajectory being synonymous with a saddle-point of $e^{iS(\gamma)}$) we could then — without in any way modifying our "integration contour" Γ — directly apply the method of stationary phase to (7) (assuming $\hbar$ to be sufficiently small). This is the familiar procedure involved in deducing a classical limit from the path integral.

It bears emphasis here that although we cannot in this manner take a classical limit within the forbidden region, there's nothing special about that region as such. The path integral (7) makes equal sense for all paths γ and for all values of the energy. What is special about having x_0 and x_1 in the forbidden region is only that we can no longer find a saddle point within the corresponding domain of integration $\Gamma(x_0, x_1)$. To obtain a useful approximation to (7) we therefore have either to find another method or to resort to an analytic continuation. As before, this can be approached in more than one way. We can "bring the saddle point to Γ", as we did above by continuing A into the complexes; or we can "bring Γ to the saddle point", as we did above by continuing z into the complexes and then deforming Γ. Let us pursue the latter avenue first, since it seems conceptually simpler than the former.

Defining $W(x) = V(x) - E$, we can write (9) more compactly as

$$S(\gamma) = \int_\gamma \frac{m}{2}\frac{dx^2}{dt} - W(x)\,dt\,, \tag{10}$$

where of course dx and dt are increments of x and t along the worldline γ. Under what circumstances will we be able to find a saddle point within our domain of integration $\Gamma(x_0, x_1)$? The condition for $\gamma \in \Gamma$ to be a saddle point is that it be a critical point of $S(\gamma)$, in other words S must be unchanged by small variations of γ. Since the limits t_i are not fixed, the most general variation of γ involves not only varying x at fixed t but also varying the range of t. The latter is equivalent to varying dt freely in (10) [or $\dot{t}$ freely in (9)], and it produces immediately the condition,

$$\frac{m}{2}\left(\frac{dx}{dt}\right)^2 + W(x) = 0\,. \tag{11}$$

Given this "energy equation", the variation of x can be ignored, since it adds almost nothing to what one obtains by differentiating (11).

Within the forbidden region, W is positive, and (11) obviously has no solution. To remedy this lack, it suffices to change the sign either of $W(x)$ or of m; but the latter looks more promising since W is a function while m is only a number. Accepting this hint, let us regard S as a function of both γ and m, and analytically continue it from positive to negative values of m. In so doing, we must however be careful that our amplitude (7) [call it $Z(m)$] remain holomorphic throughout the region through which m will pass. Otherwise it will not be possible to analytically continue Z back to positive m. But it is easy to see what this requires, given that, insofar as the term in (9) involving m is concerned, (7) is just a Gaussian integral. As long as $\Im m > 0$ the Gaussian will look like e^Q with the real part of Q being a negative-definite quadratic form in $\dot{x}$. Therefore, the Gaussian will decay at infinity and the integral defining $Z(m)$ will exist and be holomorphic in m. However m must avoid the lower half-plane because the real part of Q is positive there, and the integral would diverge. Thus, holomorphicity requires that m circle the origin in a counterclockwise sense.

When m is negative, (11) does have a solution γ, and plugging this back into (10) (or using as a shortcut the general fact that the expression $A/x +$

Bx assumes the value $\sqrt{AB}$ at its extremum), it is then easy to compute the resulting, stationary-phase approximation (4) to $Z(m)$. Or rather it is easy to compute its exponential factor, which is as far as we will carry our analysis. One obtains at the saddle-point γ, $S = -\int dx\sqrt{-2mW(x)}$ or equivalently

$$iS = -i\int dx\sqrt{-2mW(x)}\,,$$

where we of course take the positive square root of the positive number $-2mW(x)$. When we carry m back through the upper half-plane to the positive real axis, $\sqrt{-m}$ arrives as $-i\sqrt{m}$, whence iS becomes, finally

$$iS = -\int_{x_0}^{x_1} dx\sqrt{2mW(x)}\,.$$

The exponential $\exp(iS)$ then yields the familiar WKB damping factor

$$\exp\left(-\int_{x_0}^{x_1} dx\sqrt{2mW(x)}\right)\,. \tag{12}$$

Here we have obtained it via analytic continuation from a saddle-point approximation to the real-time path-integral with negative mass-parameter m.

Notice, incidentally, that we have all along been assuming that $W(x) < 0$ at both the "source" x_0 and "sink" x_1. In the contrary case where at least one of x_0 and x_1 is in the allowed region a global rotation of the sign of m (though it would still be a valid operation) would no longer secure us a saddle-point trajectory valid for all relevant x. It would be logical in that case to make m be a function of position and rotate its sign only in the forbidden region. We would then obtain a hybrid formula approximating (7) as the product of (12) from the forbidden region with the corresponding ordinary semiclassical phase-factor from the allowed region.

As conducted above, our derivation had nothing to do with imaginary time since the analytic continuation we employed touched only the mass-parameter m, leaving dt entirely alone. However, the idea of analytically continuing dt to a complex variable does arise naturally if one organizes in a certain order the infinite number of implicit integrations that a path-integral contains. Notice first that in (7) with S given by (10), the expression $d\gamma$ can be construed as an infinite product of factors $d(dx)$ and $d(dt)$, one for each "infinitesimal segment" of γ.

Equivalently, but with a symbolism that might look less confusing, we could say with reference to (9) that $d\gamma$ can be construed as an infinite product of factors $d\dot{x}$ and $d\dot{t}$. Observe moreover that the distinct increments dt (though not dx) will act as independent integration variables in our constant-energy path-integral, since it puts no constraint on their sum $t_1 - t_0$. If now we reserve the integrations over the dx for last, our path-integral resolves itself initially into a product of independent one-dimensional integrals of the form

$$\int_{dt=0}^{\infty} d(dt) e^{i\left(\frac{m}{2}\frac{dx^2}{dt} - W(x)dt\right)}, \tag{13}$$

where the limits of integration express the fact that the nonrelativistic worldline γ travels monotonically forward in time.

But we've seen this type of integral before! It's essentially the integral we used to illustrate the saddle-point method, and as we saw then, it can in some sense be evaluated exactly, depending however on the unknown value of the prefactor or "measure factor". More cautiously, we can observe that if we complexify dt then the integrand of (13) has a pair of complex saddle-points at $dt = \pm i dx \sqrt{m/2W(x)}$. Of these, the saddle point on the negative imaginary axis is the only one which is accessible in the sense in which we used that word earlier. That is, our original integration contour $(0, \infty)$ can be deformed to $-i(0, \infty)$ without changing the value of the integral. The result is the Wick rotated integral

$$-i \int_{d\tau=0}^{\infty} d(d\tau) e^{\frac{-m}{2}\frac{dx^2}{d\tau} - W(x)d\tau}. \tag{14}$$

About this Wick rotated integral we can make several comments. Firstly, although it was motivated by the existence of a saddle point for imaginary dt, the integral in itself is still exact. As such, it and its analogs in quantum field theory offer convenient starting points for Monte Carlo simulations, since the integrand is now real and positive (and formally convergent). On the other hand, one must remember that the contour-deformation that led to (14) is valid only for x in the forbidden region. Notice in this connection that the existence of a forbidden region is bound up with our having chosen to do the path-integral at constant energy E. With initial and final times held fixed instead, there is no forbidden region, as long as $V(x)$ is bounded below. where $W(x) > 0$. In the allowed region we thus should keep dt real. The corresponding saddle-point or "semiclassical path" will then have a real tangent in the allowed region and an imaginary tangent in the forbidden region.

A second comment is that we can think of the Wick-rotated path-integral is a somewhat different way that relates it more closely to our analytic continuation in the mass (see the appendix of [2]). Namely, we can introduce a complex parameter ζ into the action-integral (9) by replacing $\dot{t}$ everywhere with $\zeta\dot{t}$. When we rotate ζ from real to imaginary, the relative sign of the two terms is reversed and $iS/\hbar$ itself goes from pure imaginary to pure real, the result being essentially the same as (14). Notice that, viewed in this manner, a Wick rotation appears as a combination of two analytic continuations, one changing the sign of the mass and the other changing $\hbar$ from real to imaginary.

As illustrated by these examples, the use of imaginary time is only one device among many, that sometimes helps to simplify the approximate (or exact) evaluation of the path-integral. It has no independent status of its own beyond this. We could for example plug the saddle-point approximation to (14) back into the full path-integral (7) to obtain the latter in the form of an integral over purely spatial paths (corresponding to the so-called Jacobi action principle). If the measure-factor cooperated we could even render this "Jacobi" path-integral exact by evaluating (14) exactly. If not, it would represent a kind of cross between an exact path-integral and a saddle-point approximation to one. In either case, a further application of the saddle-point method to it, would immediately lead back to (12).

4. Tunneling in Quantum Cosmology

When it comes to cosmology, one must be cautious with the word "tunneling" since it now refers to the "birth of a cosmos from nothing". What's more, the relevant path-integral now involves a sum over 4-geometries, which seems much farther from mathematical (or even physical) respectability than what we were considering earlier. Still, if we are willing to neglect a number of important complications, we can make the problem seem remarkably close to more homely examples of tunneling, like the one treated above. This of course, is basically the view people have taken in discussing "creation of the universe via a gravitational instanton".

To simplify things as much as possible, and to make the analogy with one-dimensional tunneling as close as possible, let us consider an "empty" and (spatially) spherical cosmos which is born with zero radius and subsequently expands. Assume further that the amplitude for such a process can be expressed as an integral of the form (7), where now γ represents a spacetime geometry and $S(\gamma)$ is a suitable gravitational action, including boundary contributions as needed. In line with the spirit of this paper, I will take γ to carry a metric which is almost everywhere of Lorentzian signature. If we think of its initial portion as having the shape of a cone, then the metric cannot actually be Lorentzian at the "tip" or *origin*. Truly an origin here — not only of the coordinates, but of the cosmos. but it can still be globally smooth if it is a "Morse metric" of the type discussed in [3]. For definiteness, let us imagine the domain of the gravitational path-integral to consist of metrics of this type, in which case we can neglect any extra contribution to $S(\gamma)$ from the origin. (In 1+1 dimensions there would be such a contribution, as also in 3+1 dimensions for other topologies. See [4] and [5] for extensive discussion of this question.)

In writing down the action-integral S explicitly, we could of course employ proper time τ and spatial radius a as our coordinates, but we can lend S a much more convenient form for present purposes by employing instead the quantities T and v, where on any hypersurface Σ coinciding with a 3-sphere of homogeneity, v is the volume of Σ ($v = 2\pi^2 a^3$) and T is the spacetime volume which has accumulated up to that stage, or in other words the total 4-volume to the past of Σ ($dT = v d\tau$). (Thus, T is the time-coordinate natural to unimodular gravity.) In these coordinates the gravitational action S takes the form

$$S = \int \frac{-1}{3\kappa}\frac{dv^2}{dT} + \frac{k}{2\kappa}v^{-2/3}dT - \Lambda\, dT\,, \tag{15}$$

where $\kappa = 8\pi G$, $k = 3 \cdot 2^{5/3}\pi^{4/3} \approx 43.8$. A 3-sphere of volume v has Ricci scalar-curvature $R = kv^{-2/3}$ and we assume $\Lambda > 0$. Henceforth, I will set $\kappa = 1$.

Comparing (15) with (8) or (10) we can recognize an analogy that makes T correspond to time and v to position, while (up to sign) the cosmological constant Λ corresponds to the energy E. If we now imagine a "source" at the origin (representing perhaps the end of the previous cosmic cycle) and a "sink" at some other value of v, we have a nearly perfect analog of our tunneling problem The only thing spoiling the analogy is the sign of the kinetic term in (15), which is negative. This, of course, is an instance of the "wrong sign of conformal modes" that invalidates the usual type of Wick-rotation for the gravitational path-integral. The latter, incidentally, deforms the metric into the complexes such that $\mathrm{Imag}\,(g_{ab}) > 0$, cf. [4]. as if the "mass" were $-2/3\kappa$. If we were not restricting ourselves to spherical symmetry, this minus sign in the direction of isotropic expansion would render the overall sign of the "kinetic energy" indefinite, and this in turn would complicate any attempt to discover a valid saddle-point approximation to the path-integral. However, in our simplified situation, it suffices just to reverse the overall sign of the action S. Since this merely converts the integrand $\exp(iS)$ into its complex conjugate, it clearly will have the same effect on the "partition function" Z (assuming the "measure factor" in $d\gamma$ to be real). In setting up our analogy with tunneling, then, we can take, $m = 2/3, t = T, E = \Lambda$, and $V = k/2v^{2/3}$. (Notice also that neither 4-volume here nor elapsed time there is to be held fixed in (7), the analogy being perfect in that sense as well.)

With these substitutions, our cosmological path-integral becomes a special case of the non-relativistic path-integral we evaluated in the previous section (or rather which we approximated by various methods). Since the answer we obtained there, namely (12), was real to the accuracy to which we were working, undoing the complex conjugation doesn't change it, and we are left with $Z \approx \exp(-I)$ as the saddle-point approximation to our path-integral, where I is the integral in (12) with the above substitutions. Let us also take for v_1 the special radius (or rather volume) at which the cosmos can continue its expansion classically as a de Sitter spacetime. We can obtain this radius from the "turning point equation", $V = E$ or $\Lambda = k/2v_1^{2/3}$. The integral then yields the simple answer. (One might worry that our gravitational potential $V(v)$ is infinite at $v = 0$, but the corresponding divergence in the integrand of I is only $1/v^{1/3}$.)

$$I = \int_0^{v_1} dv \sqrt{2m(E - V)} = \frac{12\pi^2}{\Lambda}$$

and the consequent approximation

$$Z \approx \exp\left(-\frac{12\pi^2}{\Lambda}\right) . \tag{16}$$

Notice that this amplitude, if treated as the square root of a probability, would favor large values of Λ over small ones, a result that is entirely logical if one observes that small Λ corresponds to big radius at the turning point, meaning that the cosmos has to tunnel much farther during its "quantum era". Notice also that (16) expresses damping rather than amplification for a very good reason. It represents an integral over a space of Lorentzian metrics, none of which is a point of stationary phase. That in this situation the resulting cancellations lead to an exponentially small answer is just what one would expect to happen. From this point of view, arguments that have tried to associate a small Λ with an exponentially large amplitude [6,7] seem to be guilty of working with the wrong saddle-point ("instanton") of the analytically continued action. For the problem at hand, and for Λ and v_{final} fixed at real values, there seem to be a total of six such saddle-points, if I haven't miscounted. In effect, they have overlooked the conditions of validity that govern which saddle-points are accessible and which are not.

Of course any conclusion like the above remains tentative, because no one has ever (to my knowledge) embedded amplitudes like (16) in a coherent theory of quantum cosmology. One can compute some approximate amplitudes, but one doesn't know what they mean. Should one construe them as values of "the wave function of the universe", or (as I think) would it be better to conceive of them as contributions to the gravitational "quantal measure" [8], a quantity which in turn might find its ultimate interpretation in terms of notions like "preclusion" [9] and "anhomomorphic logic" [10].

Into these questions intrude other conceptual issues, which however are also more technical in nature. For example, how literally should one take the identification of our parameter T as a kind of time? To embrace that idea is to do unimodular gravity, and thence spring other questions concerning for example selfadjointness of the unimodular Hamiltonian (see e.g. [5]). Which selfadjoint extension should we choose, or, instead of selecting one, should we discard unitarity altogether, in order to provide for transitions to other topologies or to non-geometric phases? In path integral terms: How should we treat topology changing histories. And then one needs to confront the effects of anisotropy and inhomogeneity. It can happen, for example, that a saddle point which might have seemed accessible within a smaller space of highly symmetrical metrics becomes inaccessible when the constraint of symmetry is relaxed (compare e.g. the discussion of Taub vs. Friedmann in [5]).

Rather than continuing to multiply questions in this vein, let me close this section with a thought about the possible "practical" meaning of the damping-factor we have computed. Given the rather fanciful assumption of a spherical and empty cosmos, we might attach to Z in equation (16) a set of words like "the amplitude for the cosmos to arrive at size v starting from zero size". This phrase elicits a picture in which the cosmos is born "under a potential barrier" and then develops through different phases of growth. In the earliest phase, it likely has no continuum description at all. (Perhaps it has a tree-like structure [11].) Later it begins to expand as a Lorentzian manifold, but, being empty, must do so in a quantal manner by "tunneling". Finally, it emerges as a classical universe with a size dictated by the size of Λ, which (despite causal set arguments that Λ must fluctuate) I will take to be a fixed parameter. A question then is how long the cosmos spends in the first two phases. For this we can return to our nonrelativistic example where (12) governs the tunneling rate per unit time. Treating T here as our time parameter, we might thus think of (16) (or its square) as a rate of tunneling per birth of causal set element (since 4-volume simply measures number of elements). On this reading, our computation of Z would in effect have been counting the average number of births taking place during the pre-geometric and tunneling phases of cosmic growth.

5. Conclusion

Probably few people believe that clocks begin to tick off imaginary seconds the moment some alpha-particle finds itself under a potential barrier. And yet in the context of the gravitational path-integral, something like this notion has often been put forward. In opposition to such notions, a more prosaic (and more traditional) viewpoint might maintain that a tunneling amplitude, like all other quantal amplitudes, is given in the first place by a real-time path-integral. The analytically continued metrics used in connection with black hole thermodynamics present a somewhat different case than tunneling. An imaginary time-parameter arises because the operator of interest in equilibrium statistical mechanics is $\exp(-\beta\hat{H})$ rather than $\exp(-i\hat{H}t)$. One is computing a sum over stationary states, not dealing with a process developing in time. In previous sections, I have tried to illustrate a few of the different tools that are available for approximating tunneling amplitudes, and also to illustrate that some of them (like analytic continuation in the mass) are simpler in conception than those (like Wick rotation) which need to introduce a space of complexified paths γ. Thus, we considered several integrals and several different ways of treating them.

To begin with, we approximated the simple integral (2), in two different ways, first by deforming the integration contour into the complexes, and second by complexifying the real parameter A without modifying the domain of integration. With the first approach, we had to pay heed to several conditions of validity: the integrand had to continue to an analytic function; the complexified contour Γ had to be deformable back to the original contour without changing the value of the integral; Γ had to be sufficiently close to either a steepest-descent or a stationary-phase path; and the exponent $f(z)$ had to be "sufficiently rapidly varying" in the relevant sense. These conditions involved the behavior of the analytically continued integrand at infinity, and more generally some understanding of its overall "topography". Ultimately, they picked out a unique saddle point and a unique sign for the consequent approximation (4). With the second approach, related conditions of validity determined the direction of continuation of the parameter A and governed the validity of the resulting approximation.

We then turned to a path-integral for quantal tunneling, viewed as an infinite dimensional analog of the previous problem. Similar approaches were feasible and similar conditions of validity applied, the main difference being that those conditions were much more formal and much less clearly complete than in the one-dimensional case. Our first treatment of tunneling rested on complexification of the particle-mass m, which we continued from the positive to the negative real axis, in order to convert our integral $Z(m)$ into one to which the method of stationary phase would apply. We then continued the answer back to positive m and recognized our result as the familiar damping factor (12). The fact that we obtained damping rather than amplification was not arbitrary, but followed from the direction in which we continued m, and this in turn was dictated by the condition that the integral $Z(m)$ continue to be defined for the intermediate values of m. Our second treatment of the tunneling problem complexified

dt instead of m. This led to a hybrid path-integral that was "Wick rotated" in the forbidden region, but not in the allowed region. In contrast the first method could get by without treating the two regions differently, as long as both x_0 and x_1 belonged to the forbidden region. A steepest descent approximation to this integral then reproduced the same damping factor (12). Along the way, we obtained a purely spatial path-integral corresponding to the so-called Jacobi action principle. At the risk of repetition, let me emphasize that with both methods, the fact of damping as opposed to amplification is seen to be a property of the real-time path-integral itself. It emerges automatically, when one takes into account the need for consistency in the approximation method.

Finally, we carried some of this analysis over to the problem of the birth of a cosmos in quantum cosmology. Leaving to one side the thorny conceptual and technical problems that arise in this context, we wrote down a path integral describing a process in which a cosmos is born as a sphere of zero radius and expands quantum mechanically (in a "tunneling regime") until it is big enough to continue its expansion as an approximately classical (de Sitter) spacetime. This sum over Lorentzian geometries produced a problem identical in form to our earlier example of tunneling in nonrelativistic quantum mechanics. We were therefore able to solve it easily (at leading semiclassical order), and to conclude that the answer takes the form of a WKB damping factor. To the extent that the original integral is appropriate to the physics, this conclusion seems to be inescapable. In particular, it is free of any ambiguity arising from the need to choose among alternative saddle-points during the approximation process.

More generally in quantum gravity, one might expect to be able to rule certain saddle-point metrics out or in if one could pin down more of the conditions of validity that pertain to the particular situation at hand. For example reference [5] exhibited a number of saddle-point metrics that arise within unimodular quantum cosmology and considered different criteria for accepting or rejecting them. It seems that an analysis like that in the previous section could clarify further the status of some of those complex metrics and their relatives.

Acknowledgments

Thanks go to David Rideout for his help with the figures. Research at Perimeter Institute for Theoretical Physics is supported in part by the Government of Canada through NSERC and by the Province of Ontario through MRI.

References

[1] Jon Mathews and R. L. Walker, *Mathematical Methods of Physics*. Second Edition, Addison-Wesley, 1970.

[2] Rafael D. Sorkin, Forks in the Road, on the Way to Quantum Gravity, talk given at the conference entitled "Directions in General Relativity", held at College Park, Maryland, May, 1993, published in *Int. J. Th. Phys.* 36 2759–2781 (1997), arXiv:gr-qc/9706002.

[3] A. Borde, H.F. Dowker, R.S. Garcia, R.D. Sorkin and S. Surya, Causal Continuity in degenerate spacetimes, *Class. Quant. Grav.*16 3457–3481 (1999), arXiv:gr-qc/9901063.

[4] Jorma Louko and Rafael D. Sorkin, Complex Actions in two-dimensional topology change, *Class. Quant. Grav.* 14 179–203 (1997), arXiv:gr-qc/9511023.

[5] Alan Daughton, Jorma Louko and Rafael D. Sorkin, Instantons and unitarity in quantum cosmology with fixed four-volume, *Phys. Rev. D* 58 084008 (1998), arXiv:gr-qc/9805101.

[6] S.W. Hawking, The cosmological constant is probably zero, *Physics Letters B* 134 403–404 (1984).

[7] Sidney Coleman, Why there is nothing rather than something: a theory of the cosmological constant, *Nuclear Physics B* 310 643–668 (1988).

[8] Rafael D. Sorkin, Quantum Mechanics as Quantum Measure Theory, *Mod. Phys. Lett. A* 9 (No. 33) 3119–3127 (1994), arXiv:gr-qc/9401003.

[9] Robert Geroch, The Everett interpretation, Noûs 18 617–633 (1984); Rafael D. Sorkin, "Quantum Measure Theory and its Interpretation", in *Quantum Classical Correspondence: Proceedings of the* 4th *Drexel Symposium on Quantum Nonintegrability*, held Philadelphia, September 8-11, 1994, edited by D.H. Feng and B-L Hu, pages 229–251 (International Press, Cambridge Mass. 1997), arXiv:gr-qc/9507057.

[10] Rafael D. Sorkin, Quantum dynamics without the wave function, *J. Phys. A: Math. Theor.* 40 3207–3221(2007), arXiv:quant-ph/0610204.

[11] Rafael D. Sorkin, Indications of causal set cosmology, *Int. J. Theor. Ph.* 39 (7) 1731–1736 (2000). (an issue devoted to the proceedings of the Peyresq IV conference, held June-July 1999, Peyresq France), arXiv:gr-qc/0003043.

In: Recent Research in Quantum Gravity
Editor: A. Dasgupta, pp. 137-167
ISBN: 978-1-61942-386-2

Chapter 6

EUCLIDEAN QUANTUM GRAVITY: RECENT RESULTS

Arundhati Dasgupta
Department of Physics and Astronomy
University of Lethbridge
Lethbridge, Canada

Abstract

In this article a detailed description is given of the Euclidean path-integral approach to quantum gravity, and the Faddeev–Popov method of gauge fixing. The results suggest that standard methods of field theory might prove useful in obtaining a convergent path-integral evaluation of the gravitational propagator.

1. Introduction

Gravity is formulated as the theory of space-time in Einstein's theory of General Relativity. It has eluded efforts to find a 'quantum version' of the theory. Several different ways of 'quantizing' have been developed, but there is no consensus on the correct and unique way to obtain 'quantum space-time'. The reason for this confusion is the absence of any experimental verification which could confirm the theoretical predictions. In this article we review and describe new results in the *path-integral* approach to quantization which might stand the test of experimental verification. This approach to quantization has given correct results for particle mechanics and proved very successful in perturbative quantum field theory. The perturbative Feynman diagrams are one of neatest theoretical techniques which provided experimentally verifiable results tested at the accelerators. For Gravity perturbative methods give uncontrollable infinities; necessitating a search for non-perturbative methods and we will describe the non-perturbative path-integral method in this article.

The gravitational path integral is defined as the sum over all possible metrics given the boundary data, with each metric history 'weighted' with $e^{\iota S/\hbar}$ (S is the Einstein action and $\hbar$ is Plancks constant divided by 2π). This is the 'quantum amplitude' for space-time to evolve from initial boundary data to the final boundary data. Computing this has proved a difficult task, and this is the subject of this article. We review the existing approaches to the path-integral in quantum gravity and its effectiveness in obtaining the quantum amplitude for gravity. The straight forward computation of the path-integral involves an evaluation of the physical measure by factoring out the diffeomorphism group. The computation of the measure and interesting issues in gauge fixing are discussed here. A lattice regularization of the non-perturbative path-integral has succeeded in controlling infinities and obtain a emergent space-time which is fractal at short distances. A brief mention of these recent and promising developments are discussed.

In Section 2 a basic description of the path-integral and the measure is given. In section 3 the infinities which prevent the straight forward computations of the gravity amplitudes using similar techniques as in other field theories are discussed. In Section 4 the details of the computation of the measure shows a way towards a convergent path-integral. In Section 5 we discuss gauge fixing and a real computation of the path-integral as sum over histories using Lattice techniques. Section 5 is particularly important for new results, and the last section is a conclusion to the article, summarises the results and discusses the open problems.

2. The Introduction to the Gravity Path-Integral and the Faddeev–Popov Measure

The path-integral was formulated by Feynman [1] as an alternative to Schrodinger approach to quantisation. It has proved useful particularly for perturbative quantum field theory, as the computations are more transparent than its canonical formulation. The formulation is based on the following basic tenets:

1) Quantum mechanically one can predict the probability of the occurance of final events given a set of initial events.

2) There is a history of events, starting from the initial event, and each sequence is assigned a complex number or the quantum amplitude whose modulus squared is the probability. If the intermediate events in a sequence are not observed, then one sums over all plausible intermediate events.

Let us take the example of a one-dimensional particle propagating from a initial point (t_i, x_i) to a final point (t_f, x_f). The time interval is divided into a set of infinitesimal intervals each of width ϵ. The position of the particle at the time points $t_i, t_i + \epsilon, t_i + 2\epsilon, \ldots, t_i + k\epsilon$. are given by $x_1, x_2, \ldots, x_k, \ldots$. The amplitude for a particle to propagate across the infinitesimal time interval is given as a function of ϵ, as $\phi(x_{k-1}, x_k, \epsilon)$. If the intermediate positions of the particle are not observed, then one sums over the x_k. Thus the path-integral is formulated thus

$$\text{Limit}_{\epsilon \to 0} \int_{x_i t_i}^{x_f t_f} \prod_{k=1} \frac{dx_k}{A} \phi(x_{k-1}, x_k, \epsilon), \tag{1}$$

where A is a normalization constant. In the limit, this sum can be written as a integral, and the amplitude $\phi(x(t))$ emerges as a functional of the path from $x_i \to x_f$. The measure over the paths is ill defined, as there exists no mathematical proof that the limit can be taken uniquely, or even if it exists. In the limit $\epsilon \to 0$, the path integral can be written as

$$\int_{x_i}^{x_f} \mathcal{D}x \phi(x(t)). \tag{2}$$

Feynman postulated that

$$\phi(x(t)) = \exp\left(\frac{i}{\hbar}S\right),$$

where S is the action for the particle integrated from the initial point to the final point. The $\mathcal{D}x$ is the formalized measure. This path-integral, or the quantum amplitude for the particle to propagate from the initial point to another is known as the propagator $K(x_i t_i; x_f, t_f)$, and this is a function of the initial point and the final point. In case the initial time is $-\infty$, one gets the Schrodinger wave function for the particle to be

$$\psi(t, x) = \int_{-\infty}^{t} \mathcal{D}x \exp\left(\frac{i}{\hbar}S\right). \tag{3}$$

Feynman obtained the normalized path-integral measure by deriving the Schrodinger differential equation from the integral. The time is increased by a small increment of ϵ to obtain $\psi(t+\epsilon, x)$ and the Taylor series in powers of ϵ is used to obtain the differential equation for time evolution, which turns out to be the Schrodinger equation for the wavefunction. To get the exact Schrodinger equation, the normalisation in the measure is identified as

$$\mathcal{D}x = \text{Limit}_{\epsilon\to 0} \prod \frac{dx_{\epsilon_k}}{\left(\frac{ih\epsilon}{m}\right)^{1/2}} \tag{4}$$

mass of the particle being m. The analytic continuation in time takes $\epsilon \to i\epsilon$, and the measure to a set of real numbers, and this defined the Weiner measure. Away from the saddle point, the paths which contribute most to the path-integral are non-classical paths. These are piecewise analytic in the ϵ interval, but have sharp corners. These paths are drawn in green in the Figure (1). In the $\hbar \to 0$ limit, the oscillatory factor in the exponent contributes maximum when the action varies least rapidly, or its variation with respect to the path vanishes. This saddle point approximation is discussed in details in Chapter 5; article by R. Sorkin in this volume.

Away from the saddle point the integral for the free particle or the Harmonic oscillator can be computed as the integral is Gaussian. For particles in other potentials one can do a perturbation about the Gaussian and this is the technique

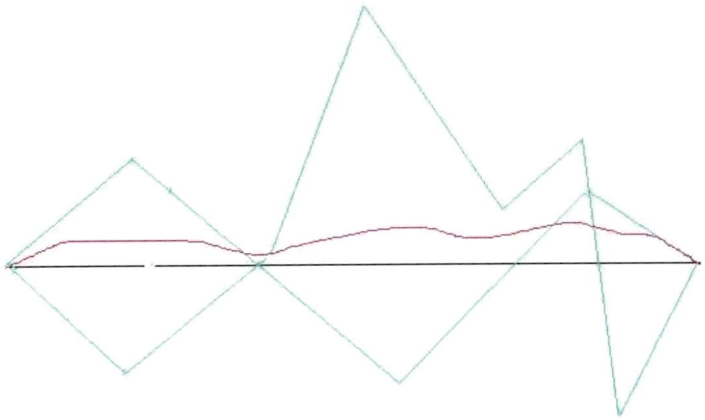

Figure 1. Non-classical Paths.

which was easily extended to computations in field theory using the Feynman diagrams [9].

The continuation of the time coordinate to the imaginary axis, transformed the complex integrand to a real Boltzmann weight. The measure in the 'Euclidean space' could be mathematically defined, and the Euclidean path-integral is

$$\int \mathcal{D}_E x \exp\left(-\frac{S_E}{\hbar}\right). \tag{5}$$

Eventually one computes the Euclidean version and rotates back to Lorentzian signature in the final answer. In this article this 'Euclidean path-integral' is the subject of discussion. In exact analogy with Feynman's initial formulation a path-integral for a space-time defined using the boundary metric $h^1_{\mu\nu}$ defined on a initial surface Σ_i can evolve into another to boundary metric $h^2_{\mu\nu}$ defined on a final surface Σ_f is given by

$$K(h^1_{\mu\nu}\Sigma_1, h^2_{\mu\nu}\Sigma_2) = \int \mathcal{D}g_{\mu\nu} \exp\left(\frac{i}{\hbar}\int d^4x\sqrt{\det g}\, R\right), \tag{6}$$

where the Einstein's action defined by the determinant $\det g$ and the scalar Ricci Curvature R of all metrics appears in the exponent. There is considerable discussion about whether one can take a parameter like time to describe the flow from one boundary to another, as diffeomorphism invariance obliterates coordinate time as a physical observable or an external parameter. However the path-integral can be formally defined given any boundary data, despite the simplicity of its definition, it has proven very difficult to compute as

(i) The gravitational action is diffeomorphism invariant, and hence geometries with the same action are over counted in the measure.

(ii) Analytic continuation in time does not make sense as 'time' coordinate is not uniquely identified in a diffeomorphism invariant theory.

(iii) The gravitational action is non-quadratic and even non-polynomial due to the $\sqrt{\det g}$ term, and we can compute integrals which are essentially Gaussian.

(iv) Perturbation about a classical metric with well defined time and metric does not make sense as Feynman diagrams diverge at each order in the perturbation theory and the theory is non-renormalisable [2, 3].

(v) If one formulates a Euclidean path-integral ab-initio, one finds that the gravitational path-integral is unbounded from below, and the exponent or the probability for certain configurations diverges making the 'integral' including those configurations meaningless [4].

Several physicists have been working to solve these issues, or formulate quantum gravity using new approaches including

(a) Causal Dynamical Triangulation (CDT) pioneered by Ambjorn, Jurkiewicz and Loll [5].

(b) Spin Foam approach which is a sum over histories computation of the Loop Quantum Gravity spin-networks [6].

However, in this article we shall see that even at the level of a traditional continuum evaluation, there can be interesting observations and the gravitational path-integral is not as badly divergent as previously thought. The path-integral computation begins with a evaluation of the measure as the ' physical measure' has to be computed by factoring out the diffeomorphism group. This problem does not exist in the point particle computation as the paths which contribute to the measure are not related by any symmetry of the theory. To begin, I discuss the issue of the computation of the physical measure.

2.1. Faddeev–Popov Measure

The gravitational action as formulated by Einstein $S = \frac{1}{16\pi G}\int\sqrt{\det g}\, R$ (one could include a cosmological constant term proportional to Λ) is invariant under transformations of the metric $g'_{\mu\nu} = \frac{\partial x^\lambda}{\partial x'^\mu}\frac{\partial x^\rho}{\partial x'^\nu} g_{\lambda\rho}$, where a coordinate transformation $x^\mu \to x'^\mu$ is implemented. To find the generators of the transformation, one takes a infinitesimal coordinate transformation, implemented thus $x^\mu \to x^\mu + \xi^\mu(x^\mu)$. The ξ^μ are the vector fields, the generators of the diffeomorphisms. Thus configurations which are related by coordinate transformations leave the action invariant, and represent redundant or unphysical degrees of freedom. The measure however counts them as distinct geometries, and one has to factor out the diffeomorphism group. The path-integral written in terms of physical degrees of freedom is thus

$$\int \frac{\mathcal{D}g_{\mu\nu}}{Diff(M)} \exp\left(\frac{i}{\hbar}S\right), \tag{7}$$

where $Diff(M)$ is the diffeomorphism group of the underlying manifold M. Apart from the diffeomorphism group, there also exists another set of transformations of the metric, the 'conformal transformations'.

$$g_{\mu\nu} = \exp(2\phi)\bar{g}_{\mu\nu} \tag{8}$$

these of course do not leave the action invariant (except for conformal Killing directions). We shall talk about the conformal mode ϕ and the conformal equivalence class of metrics $\bar{g}_{\mu\nu}$ in the subsequent discussions. To isolate the physical measure, it is always useful to talk about De-Witt superspace, which is a manifold comprising of points each of which is a metric. Taking one particular slice of this manifold, identifies a unique set of metrics, and rest of the superspace can be obtained by using the diffeomorphism or conformal orbits. Thus, in a infinitesimal neighbourhood of the slice, using the cotangent space element $\delta g_{\mu\nu} = h_{\mu\nu}$, one obtains [7]

$$h_{\mu\nu} = h^{\perp}_{\mu\nu} + \nabla_\mu\xi_\nu + \nabla_\nu\xi_\mu + 2\phi g_{\mu\nu}. \tag{9}$$

The $h^{\perp}$ represents the traceless tensor modes on the gauge slice, or where the conformal mode has been factored out and written isolated as the third term.

This is further written in a more useful way

$$\begin{aligned} h_{\mu\nu} &= h^{\perp}_{\mu\nu} + \left(\nabla_\mu \xi_\nu + \nabla_\nu \xi_\mu - \frac{1}{2}\nabla.\xi g_{\mu\nu}\right) + \left(2\sigma + \frac{1}{2}\nabla.\xi\right) g_{\mu\nu} \\ &= h^{\perp}_{\mu\nu} + (L\xi)_{\mu\nu} + \left(2\phi + \frac{1}{2}\nabla.\xi\right) g_{\mu\nu}, \end{aligned} \quad (10)$$

where clearly, the trace of the diffeomorphisms is isolated as it contributes to the conformal sector of the theory and a operator $L\xi$ defined which maps vectors to traceless two tensors. The implementation of this coordinate transformation in the cotangent space of metrics leads to the identification of a Jacobian, which is the Faddeev–Popov determinant. In the space of metrics, the coordinate transformation also leads to the same Jacobian. Thus the measure in the path-integral is written in terms of the physical metrics and the Jacobian, and the diffeomorphism group is factored out. The path integral looks thus:

$$\int \mathcal{D}g^{\perp}_{\mu\nu}\mathcal{D}\phi(\mathrm{DetM}) \times \exp\left(\frac{\iota}{16\pi G\hbar}\int d^4x\,\sqrt{g^{\perp}}\,e^{2\phi}\left[R^{\perp} + 6(\nabla\phi)^2\right]\right). \quad (11)$$

The Faddeev–Popov determinant can be calculated formally as a functional determinant, however, evaluating the determinant as a function of the physical metric is a task which requires regularization and appropriate use of new techniques. Perturbative evaluations give infinities. A non-perturbative evaluation of the determinant exists for the scalar sector of the determinant [8]. The interesting factorization of the Faddeev–Popov determinant occurs into a scalar determinant and the vector and tensor ones by using a redefinition of the diffeomorphisms. The vector ξ which generates the diffeomorphism is broken up into a divergence less vector $\hat{\xi}$ and a divergence of a 'scalar' σ

$$\xi_\mu = \hat{\xi}_\mu + \nabla_\mu \sigma, \quad (12)$$

$$\nabla^\mu \xi_\mu = \nabla^\mu \nabla_\mu \sigma. \quad (13)$$

Thus the trace of the diffeomorphism $\nabla.\xi$ is a function of the scalar σ. .

3. The Infinities

In this section we discuss the divergences which occur in the path-integral which compells us to search for the non-perturbative techniques, and compute the physical measure. The first is the unboundedness of the Euclidean Einstein action, and second the perturbative infinities which arise when one actually starts computing the gravitational path-integral.

3.1. The Conformal Mode and the Unboundedness of the Euclidean Action

The Euclidean gravitational action or the action for positive definite metric comprises of (unless otherwise stated the metric and the action from this point will comprise of Euclidean metrics)

$$S = -\frac{1}{16\pi G}\int \sqrt{\det g}\, R\, d^4x. \tag{1}$$

The analytic continuation from Lorentzian metrics and the nature of the contour determines the negative sign. In [4] Hawking showed that if one isolates the conformal mode of the metric (8) it appears in the Euclidean action thus;

$$S = -\frac{1}{16\pi G}\int d^4x\, e^{2\phi}\sqrt{\bar{g}}\left[\bar{R} + 6(\bar{\nabla}\phi)^2\right]. \tag{2}$$

The kinetic term of the conformal mode is positive definite, and hence the Euclidean action can assume as negative values as possible. This unboundedness of the Euclidean action makes the Boltzmann weight in the Euclidean path-integral infinite for certain metrics. Thus even before the sum over histories can be computed there are 'singular configurations' which have to be 'avoided' in order to make the integral computable.

This pathology can be assumed to be a signature of presence of redundant degrees of freedom, and indeed diffeomorphically related metrics in the measure are redundant. So, in this paper we investigate the path-integral written in terms of physical variables, which we have already specified means a computation of the Faddeev–Popov determinant.

Does the conformal mode however uniquely isolate any divergence in the Euclidean action? The $\sqrt{\bar{g}}e^{2\phi}\bar{R}$ term in the action has the 'kinetic terms' of the conformally equivalence class of metrics. For compact manifolds we can use the Yamabe Conjecture to reduce the $\bar{R}$ to a constant, and hence this term cannot make the action unbounded by itself, and thus we ignore the contribution. (Yamabe conjecture states that any metric on a compact manifold can be conformally transformed to a metric which has constant scalar curvature, positive, negative or zero.) In case of arbitrary manifolds with boundaries, like the case of flat space time, a slightly more careful treatment is required. We take the form of the Ricci scalar given in terms of the derivatives of the metric

$$\sqrt{\bar{g}}\bar{R} = \sqrt{\bar{g}}\Big[\bar{g}^{\mu\sigma}\bar{g}^{\nu\rho}\partial_\mu\partial_\nu\bar{g}_{\sigma\rho} - \bar{g}^{\mu\nu}\bar{g}^{\alpha\beta}\partial_\mu\partial_\nu\bar{g}_{\alpha\beta}\Big]. \tag{3}$$

In the above the second term is potentially problematic, to analyse the terms in a algebraic way, we rewrite the second term in terms of the derivative of the determinant of the metric $\bar{g}$. This makes the action take the form:

$$\begin{aligned}\sqrt{\bar{g}}\bar{R} &= \sqrt{\bar{g}}\Big[\bar{g}^{\beta\nu}\bar{g}^{\alpha\sigma}\partial_\beta\partial_\alpha g_{\sigma\nu} \\ &\quad + \frac{1}{\bar{g}^2}\bar{g}^{\alpha\beta}\partial_\alpha\bar{g}\partial_\beta\bar{g} + \bar{g}^{\alpha\beta}\partial_\alpha\bar{g}_{\mu\nu}\partial_\beta\bar{g}^{\mu\nu} - \frac{\bar{g}^{\alpha\beta}}{\bar{g}}\partial_{\alpha\beta}\bar{g}\Big].\end{aligned} \tag{4}$$

Thus a potential divergent contribution to the action might come from the second term of the above for rapidly varying $\bar{g}$ (the other terms in the above donot contribute with a unique sign, and thus in the action integral negative contributions will cancel positive contributions instead of summing up to give unbounded behavior). In case of non-compact geometries we might take the $\bar{g}_{\mu\nu}$ to be unimodular and then the divergence of the Euclidean action will be concentrated in the conformal mode term. In case of arbitrary $\bar{g}_{\mu\nu}$, the term has the same sign as the conformal mode term, and thus as discussed in this article, any remnant divergence from the $\bar{g}_{\mu\nu}$ can be cancelled by the contribution from the measure. However, a more careful analysis of the divergent terms in the Euclidean action are required. For the purposes of this article we will see that the solution given cures problems due to any divergent Euclidean action. This as evident from the previous discussion is due to rapidly varying volume ($\bar{g}$) of any space-time.

3.2. The Perturbative Infinities

It is very well known that the perturbative path-integral is non-renormalisable. This basically means that the usual Feynman diagram technique of perturbatively evaluating quantum amplitudes regulates infinities by extracting the finite part in other field theories, but in gravity the infinities are infinite and cannot be regulated. The computation of the path-integral for gravity by perturbative methods yields infinite terms at every order in the loop diagrams. This can be seen from easy superficial counting of the divergent diagrams [9].

The Feynman diagrams are a technique, prescription for computing quantum amplitudes as a series in the coupling strengths of the interaction terms [9]. In gravity due to the non-linearity of the action, there are self interaction terms. In parturbative gravity, one takes the metric as a fluctuation over flat space-time, or in the Euclidean case, simply $g_{\mu\nu} = \delta_{\mu\nu} + h_{\mu\nu}$. The quantum amplitude is a sum over the $h_{\mu\nu}$ field. In a particular gauge the gravitational classical action $\sqrt{\det(\delta_{\mu\nu} + h_{\mu\nu})}\,[R(h_{\mu\nu})]$ yields infinite number of vertices or interaction terms. The perturbative gravity action is using (3)

$$\begin{aligned}\int d^4x\sqrt{\det(1+h)}\\ \times\Big[(\delta^{\beta\nu} - h^{\beta\nu} - h^{\beta\tau}h^{\nu}_{\tau} + O(h^3))(\delta^{\alpha\sigma} - h^{\alpha\sigma} - h^{\alpha\tau}h^{\sigma}_{\tau})h_{\sigma\nu,\beta\alpha}\\ - (\delta^{\beta\nu} - h^{\beta\nu} - h^{\beta\tau}h^{\nu}_{\tau})(\delta^{\alpha\sigma} - h^{\alpha\sigma} - h^{\alpha\tau}h^{\sigma}_{\tau})h_{\alpha\sigma,\beta\nu}\Big].\end{aligned} \tag{5}$$

It is clear that there are a infinite number of interaction terms of the $h_{\mu\nu}$ field if one kept the series to all orders of the inverse $h^{\mu\nu}$ and the square root of the determinant. The quadratic terms in the lagrangian can be obtained by completing the square, and the propagator computed in a particular gauge. The graviton propagator in De-donder gauge $\partial_\mu h^{\mu\nu} = 0$ is computed using the standard methods of quantum field theory [9]

$$(\delta_{\mu\sigma}\delta_{\nu\tau} + \delta_{\mu\tau}\delta_{\nu\sigma} - \delta_{\mu\nu}\delta_{\sigma\tau})/p^2, \tag{6}$$

where $p^2 = p_\mu p^\mu$ is the graviton momentum. For the vertices, which are of n order, the vertex functions become increasingly complicated but it is obvious that all the vertices are functions of two powers of momenta (there are two deritives of the h). Thus if one does a superficial divergence power counting for the Feynman diagrams then at each order that number D is [2]

$$D = -2L_i + 2\sum_n V_n + 4k \tag{7}$$

Figure 2. Feynman Loop Corrections.

The L_i are the number of internal lines of the diagram (each contributes with two powers of momenta in the denominator), V_n is the nth order vertex, each contributing with two powers of momenta in the numerator, and the number of independent momenta k contributes with 4 positive powers of momenta in the integral measure. This counting usually gauges the divergence of the loop-inegral. If $D > 0$ it means that the integral is divergent, if $D < 0$ it means that the integral is convergent. It also happens that as the number of independent momenta $k = L_i - \sum_n V_n + 1$ ($L_i$is the number of internal lines - the conservation equations at each vertex V_n + one free from the conservation equations). Thus the

$$D = 2(k+1). \tag{8}$$

This is a problematic result and tells us that the gravitational divergence increases with the order of the diagramm. Thus to cancel the divergences one has to subtract terms at each order in the perturbation theory implying that the theory is non-renormalisable. In Figure 2, diagram (A) $L_i = 2$, $V_3 = 2$, $K = 1$; $D = 4$ in diagram (B) $L_i = 4$, $V_3 = 2$, $V_4 = 1$, $K = 2$, $D = 6$. This is in contrast to Yang–Mills theory, which has a three vertex with linear function of momenta and a four vertex which is independent of momentum. The superficial degree of divergence for Yang–Mills theory is $D = 4 - L_e$ where L_e is the number of external legs. If the Feynman diagrams in 2 were of gluons, the D would be 2 for both (A) and (B). Thus adding counter terms of dimension 2 to the bare Yang–Mills Lagrangian would cancel the divergence for each diagram.

4. Using The Measure in the Path-Integral

We now calculate the Faddeev–Popov determinant, but using (12). The interesting aspect of this new calculation is the separation of the trace of the diffeomorphisms and evaluation of the Faddeev–Popov determinant for this. The path-integral is defined to be

$$Z = \int \mathcal{D}\phi \, \mathcal{D}\bar{g}_{\mu\nu} \exp\left(\frac{1}{16\pi G}\int d^4x e^{2\phi} \sqrt{\bar{g}}[\bar{R} + 6(\bar{\nabla}\phi)^2]\right) \tag{1}$$

(we subsequently set $16\pi G$=1) To identify the correct measure one writes in the cotangent space of the De-Witt super space [7] or Equation (10), a coordinate transformation to the traceless gauge fixed part (denoted by a $\perp$) and a trace part and the pure diffeomorphisms generated by $\xi_\mu = \hat{\xi}_\mu + \nabla_\mu \sigma$. The Jacobian of the coordinate transformations is the Faddeev–Popov determinant and written as (det M) subsequently.

$$Z = \int \mathcal{D}\phi \, \mathcal{D}g^{\perp}_{\mu\nu} \, \mathcal{D}\hat{\xi}_\mu \, \mathcal{D}\sigma \det M \\ \times \exp\left(\int d^4x \sqrt{g^{\perp}} \, e^{2\phi}[R^{\perp} + 6(\bar{\nabla}\phi)^2]\right). \tag{2}$$

This way of gauge fixing is completely non-perturbative, and is not specific to a gauge fixing. The det M is shown to have a scalar determinant [8] times a vector and a tensor determinant.

$$\det M = \left[\det{}_S(-\nabla^2)\right]^{-1/2} \det{}_S \\ \times \left[8(1+2C)\Big(-2(\nabla)^4 + 4\nabla_\mu\nabla^2\nabla^\mu + 4\nabla_\mu\nabla_\rho\nabla^\mu\nabla^\rho\Big)\right]^{1/2} \\ \det{}_V\tilde{V} \det{}_T T. \tag{3}$$

The first determinant appears due to the Jacobian picked up due to a transformation (12). The operators V_μ and $T_{\mu\nu}$ which are vector and tensor operators can be easily identified as given in the [8]. ∇^μ is the covariant derivative operator and ∇^2 is the Laplacian with respect to the metric $g_{\mu\nu}$. (In case the derivative operators are evaluated for $\bar{g}_{\mu\nu}$ they are represented by $\bar{\nabla}$ operators.) C is the constant in the De-Witt metric, the metric of superspace which determines its signature (see [8, 11] for definition). The De-Witt metric is of indefinite signature for $C < -1/2$, and Einstein gravity has $C = -2$. The gauge volume or the integrations over the $\hat{\xi}_\mu$ and the σ can be taken out of the path-integral, and the Faddeev–Popov determinant remains to contribute to the effective action. Note that since this isolation of the trace is independent of the gauge fixing condition, our results will be true in any gauge.

4.1. Perturbative Case

We begin by taking the perturbative case and give the resolution of the conformal mode problem in some known cases using the measure in (3). In the perturbative situation ϕ is taken to be very small, thus $e^{2\phi} = 1 + 2\phi + 2\phi^2$ and $\bar{R}$ is the curvature of the space-time one is perturbing about; hence the action gives

$$
\begin{aligned}
S &= -\int d^4x\sqrt{\bar{g}}(1+2\phi+2\phi^2)\left[\bar{R}+6(\bar{\nabla}\phi)^2\right] \\
&= -\int d^4x\sqrt{\bar{g}}\left[6(\bar{\nabla}\phi)^2+2\phi\bar{R}+2\phi^2\bar{R}+\bar{R}\right] \\
&= -\int d^4x\sqrt{\bar{g}}\left\{2\phi'\left[-3\bar{\nabla}^2+\bar{R}\right]\phi'+\bar{R}-\frac{1}{2}\,\bar{R}(\Delta^{-1})\bar{R}\right\}.
\end{aligned}
$$

Where the square in $\phi' = \phi + \frac{1}{2}\,\Delta^{-1}\bar{R}$ is completed and one obtains a non-local term in that process ($\Delta \equiv -3\bar{\nabla}^2 + \bar{R}$).

For perturbations about Minkowski space-time, $R_{\mu\nu\lambda\rho} = 0$ in the first approximation, and one obtains the scalar determinant (3) as

$$
\left[\det_S(-\nabla^2)\right]^{-1/2}\det_S\left[8(1+2C)(6\bar{\nabla}^4)\right]^{1/2}. \tag{4}
$$

Under suitable boundary conditions, the determinant splits into

$$
\det_S(-\nabla^2)^{-1/2}\det_S\left[-8(1+2C)6\bar{\nabla}^2\right]^{1/2}\det_S[-\bar{\nabla}^2]^{1/2}. \tag{5}
$$

The second order operator has divergent determinant for $C < -1/2$, which is the regime of Einstein gravity. The zero modes of the above determinants and the conformal killing directions, have to be factored to give a meaningful answer [10]. For the purposes of this paper, we assume that they have been factored as required. Thus the partition function acquires the following form in the conformal sector:

$$\int \mathcal{D}\phi \det_S \left[-48(1+2C)\bar{\nabla}^2\right]^{1/2} \exp\left(-\int d^4x\sqrt{g}\,\phi'[6\bar{\nabla}^2]\phi'\right). \tag{6}$$

Clearly the ϕ integral can be written formally as a determinant, which is divergent.

$$\det{}_S[-48(1+2C)\bar{\nabla}^2]^{1/2}\frac{1}{\det_S[6\bar{\nabla}^2]^{1/2}}. \tag{7}$$

Thus the divergent determinants from the Faddeev–Popov and the ϕ integral can be cancelled for $C < -1/2$. The remaining terms after the cancellation are convergent and will give finite answers for the path-integral.

Next the De-Sitter and anti De Sitter backgrounds which are constant curvature metrics and hence have $\bar{R}_{\mu\nu} = \Lambda g_{\mu\nu}$ are considered (Λ is a cosmological constant), and the conformal mode is treated as a fluctuation over that.

The scalar determinant (3) factorizes rather neatly as fourth order operator is

$$6\nabla^4 + 4[\nabla_\mu, \nabla^2]\nabla^\mu + 4\nabla_\mu[\nabla_\rho, \nabla^\mu]\nabla^\rho = 6\nabla^4 + 8\Lambda\nabla^2. \tag{8}$$

The determinant can be factorised for the above operator using $\bar{R} = 4\Lambda$, as

$$\begin{aligned}\det_S(-\bar{\nabla}^2)&\left[8(1+2C)(-3\bar{\nabla}^2-\bar{R})\right]\\ &= \det_S(-\bar{\nabla}^2)\det_S\left[8(1+2C)(-3\bar{\nabla}^2-\bar{R})\right].\end{aligned} \tag{9}$$

The action for the scale factor in the De-Sitter or anti De-Sitter backgrounds is

$$\begin{aligned}\int d^4x\sqrt{\bar{g}}(1+2\phi+2\phi^2)&\left[\bar{R}+6(\bar{\nabla}\phi)^2\right]\\ &+\int d^4x\sqrt{\bar{g}}(1+4\phi+8\phi^2)(-2\Lambda).\end{aligned} \tag{10}$$

When the square is completed one obtains $\int d^4x\sqrt{\bar{g}}\,2\phi'(-3\bar{\nabla}^2 - \bar{R})\phi'$, where we have put in $\bar{R} = 4\Lambda$. The integral over ϕ' thus gives a divergent determinant which is inverse to second of the factored determinants of (9) and is thus cancelled. Thus the partition function for pure conformal fluctuations about De Sitter space is given by N which is a normalization, but clearly the conformal mode divergence has been cancelled. So, for most Ricci flat and constant curvature metrics, we seem to have correctly identified the resolution. The zero modes of the operator still remain. This resolution is similar to that described in [10]. There remain the transverse fluctuations of the metric, however which is beyond the scope of the discussion of the conformal mode.

4.2. Non-perturbative Measure

For the non-perturbative case, no such neat factorizations of the determinant occur, and one has to compute the functional determinant using known techniques like the heat kernel equation. Ab initio, the evaluation of the determinant (3) in the non-perturbative regime appears very difficult, as the operator is a fourth order differential operator. The scalar operator (3) is clearly a self adjoint operator (the ∇^4 term is obviously self adjoint and the $\nabla_\mu R_{\mu\nu}\nabla_\nu$ is also self adjoint due to the symmetry of $R_{\mu\nu}$ under the interchange of the μ and the ν indices). But we can try to obtain the determinant using Heat Kernel techniques [11].

The heat kernel is defined in order to achieve a ζ function regularization of the determinant. Given an operator F with eigenvalues λ, the zeta function of p is defined to be

$$\zeta(p) = \sum_{\lambda} \frac{1}{\lambda^p}. \tag{11}$$

Clearly,as the determinant of the operator is given by

$$\det(F) = e^{\mathrm{TrlnF}} \tag{12}$$

and thus

$$\det(F) = e^{-\zeta'(0)}. \tag{13}$$

Other regularization schemes can also be used if required, thus the exact answer would be particular to the way of regularization. From (13), the Faddeev–Popov determinant can be written as an exponential and adds to the classical action in the 'weight' of the path-integral creating an effective action. Since the Faddeev–Popov has the square root of the determinant appearing in the measure, the exact terms which appear in the effective action for the scalar determinant is $\Gamma_{\mathrm{trace}} = \frac{1}{2}\zeta'(0)$. Thus one has to find $\zeta'(0)$ for the scalar determinant of (3).

The zeta function for a given operator is appropriately written in terms of the 'heat kernel'.

$$\zeta(p) = \frac{\mu^{2p}}{\Gamma(p)} \int dt\, t^{p-1} \mathrm{Tr} \exp(-tF)(x, x'). \tag{14}$$

The heat kernel is precisely the term $U(t, x, x') = \exp(-tF)(x, x')$, t is a parameter and x, x' represent coordinates of the manifold in which the operator is defined. The heat kernel also satisfies the differential equation

$$\left(\frac{\partial}{\partial t} + F\right) U(t, x, x') = \delta(x, x')\big|_{t=0}, \tag{15}$$

where the boundary condition is that of a diffusion equation. The heat kernel can be solved exactly for the Laplacian in flat space, and for curved space-time, De-Witt wrote a particular ansatz, which has been solved partially. The ansatz for the heat kernel for the Laplacian is an expansion in powers of the parameter t

$$U(t, x, x') = \frac{1}{(4\pi t)^2} e^{-\frac{\bar{\sigma}}{2t}} \sum_n a_n(x, x') t^n, \tag{16}$$

where $\bar{\sigma}$ is half of the square of the geodesic distance existing given two points (x, x'). The actual evaluation of the heat kernel and the determinant of the operator is done in the coincidence limit $x \to x'$.

The procedure for finding the heat kernel for the scalar operator which appears in the determinant in the measure is simplified slightly, by solving in two different regimes. The fourth order scalar operator in (3) is taken thus

$$-2\nabla^4 + 4\nabla_\mu \nabla^2 \nabla^\mu + 4\nabla^\mu \nabla^\rho \nabla_\mu \nabla_\rho. \tag{17}$$

By using the commutation relations, one obtains

$$6\nabla^4 + 4\nabla_\mu [\nabla^2, \nabla^\mu] + 4\nabla^\mu [\nabla^\rho, \nabla_\mu] \nabla_\rho \tag{18}$$

$$6\nabla^4 + 8\nabla_\mu R^{\mu\nu} \nabla^\nu. \tag{19}$$

We take two limits, one where there is weak field gravity, and one obtains $\nabla_\mu R^{\mu\nu} \sim 0$, here (19) is approximated by

$$6\nabla^4 \left(1 + \frac{4}{3} (\nabla^4)^{-1} \nabla_\mu R^{\mu\nu} \nabla_\nu\right). \tag{20}$$

And where gravity is strong, one obtains

$$8\nabla_\mu R^{\mu\nu}\nabla_\nu\left(1+\frac{3}{4}\left(\nabla_\mu R^{\mu\nu}\nabla_\nu\right)^{-1}\nabla^4\right). \tag{21}$$

This particular approximation does not appear in any previous calculation, usually the expansion of higher order operators has been done in orders of $(\nabla^2)^{-1}$ *[12].* Thus for the purposes of the kinetic term of the conformal mode, which is the reason for the unboundedness of the Euclidean action, it is enough to obtain the determinant of the following scalar operators in the above two limits.

$$\det{}_S\left(8(1+2C)6\nabla^4\right) \quad (\nabla^\mu R_{\mu\nu}\sim 0) \tag{22}$$

and

$$\det{}_S\,(8(1+2C)8\nabla_\mu R^{\mu\nu}\nabla_\nu) \quad (\nabla^\mu R_{\mu\nu}\gg 1). \tag{23}$$

The first one (22) can be evaluated using the Heat Kernel for the Laplacian for arbitrary space times, under certain boundary conditions, by splitting the fourth order operator into product of two Laplacians whose heat kernel expansions are well known [11].

The (23) operator is clearly non-Laplacian type and non-minimal in the sense that $R_{\mu\nu}$ is not covariantly constant. Thus, I use a new ansatz for the heat kernel, where

$$U(t,x,x') = \frac{1}{(4\pi t)^2}\exp(-\varepsilon/2t)\sum_n a_n t^n, \tag{24}$$

where ε, is a generalization of $\bar{\sigma}$ used by De-Witt, and is determined by solution to an equation written below. One then finds a recursion relation for the a_i using the differential equation for the definition of the heat kernel (15).

The heat kernel differential equation (15) for $t \neq 0$ is written for the operator $(-8\nabla^\mu R_{\mu\nu}\nabla^\nu)$. We use a negative sign in front of the determinant as eventually the operator whose determinant will be used has $C = -2$ in the prefactor which is negative.

$$\left(\frac{\partial}{\partial t} - 8\nabla_\mu R^{\mu\nu}\nabla_\nu\right)\frac{1}{(4\pi t)^2}\exp(-\varepsilon/2t)\sum a_n t^n = 0,$$

or

$$\left(\frac{\partial}{\partial t} - 8\nabla^\mu R^{\mu\nu}\nabla^\nu - 8R^{\mu\nu}\nabla^\mu\nabla^\nu\right)\frac{1}{(4\pi t)^2}\exp(-\varepsilon/2t)\sum a_n t^n = 0.$$

The linear term can be removed by a scaling by $\exp(\bar{\Lambda}(x))$. The function $\bar{\Lambda}$ is determined according to the differential equation $\nabla_\tau(\bar{\Lambda}) = -\frac{1}{2}(R^{-1})_{\nu\tau}\nabla_\lambda R^{\lambda\nu}$. This reduces the above equation to

$$\left(\frac{\partial}{\partial t} - 8R_{\mu\nu}\nabla^\mu\nabla^\nu + Q\right)\frac{1}{(4\pi t)^2}\exp(-\varepsilon/2t)\sum a_n t^n = 0, \quad (25)$$

where a potential $Q = -8\nabla_\mu R^{\mu\nu}\nabla_\nu\bar{\Lambda} - 8R^{\mu\nu}\nabla_\mu\nabla_\nu\bar{\Lambda}$ gets added due to the scaling of the wavefunction. The recursion relations for coefficients a_n obtained from (25) are:

$$\frac{\varepsilon}{2} - 2R_{\mu\nu}\nabla^\mu\varepsilon\nabla^\nu\varepsilon = 0, \quad (26)$$

$$8R_{\mu\nu}\nabla^\mu\nabla^\nu a_{n-1} - 8R_{\mu\nu}\nabla^\nu\varepsilon\nabla^\mu a_n - 4\nabla^\mu\nabla^\nu\varepsilon R_{\mu\nu}a_n$$
$$-(n-2)a_n - Qa_{n-1} = 0. \quad (27)$$

The obvious solution for (26) is to take $\frac{1}{4}\,\varepsilon = R_{\mu\nu}\nabla^\mu\varepsilon\nabla^\nu\varepsilon$. This equation is solved easily in the case where the Ricci curvature is constant $R_{\mu\nu} = \frac{1}{l^2}\,g_{\mu\nu}$, and using that a generalization is given for arbitrary curvature metrics in Riemann normal coordinates.

The case of the constant curvature (25) is solved by:

$$\frac{\varepsilon}{4} = \frac{1}{l^2}\,g_{\mu\nu}\nabla^\mu\varepsilon\nabla^\nu\varepsilon, \quad (28)$$

$$\varepsilon = \frac{l^2}{8}\,\bar{\sigma}, \quad (29)$$

where $\bar{\sigma}$ is half of the square of the geodesic distance from the point x to x'.

The case of arbitrary curvature, the Riemann normal coordinate expansion is taken for the metric and a solution is given in the coincidence limit only $x \to x'$. The Riemann normal coordinates are used to expand for the metric and the curvature, about a given point, similar to a Taylor series for a function. The standard form for the square of the geodesic distances is

$$2\bar{\sigma}(x, x') = g_{\alpha\beta}\bar{\sigma}^\alpha\bar{\sigma}^\beta, \quad (30)$$

where $\bar{\sigma}^{\alpha}(x, x')$ is the tangent vector to the geodesic at the point (x). In Riemann Normal coordinates, this is

$$2\bar{\sigma}(x, x') = \delta_{ab}\hat{x}^a\hat{x}^b, \tag{31}$$

where $\hat{x}^a$ is the coordinate joining the origin of the Riemann normal coordinates to a nearby point, or $(x - x')^a$ when x is taken as the origin and $x' \to x$. Motivated from equation (29), the ε can be written in Riemann normal coordinates as:

$$\varepsilon = \frac{1}{16} g_{ab}(R^{-1/2})^{ac}\hat{x}_c(R^{-1/2})^{bd}\hat{x}_d. \tag{32}$$

This has the interesting properties in the coincidence limit (though there are no obvious interpretations for ε as there are for σ in terms of geodesics and tangents at the points x, x')

$$\nabla^a\varepsilon = 0, \tag{33}$$

$$\nabla^a\nabla^b\varepsilon = \frac{1}{8}(R^{-1})^{ab}, \tag{34}$$

$$\nabla^a\nabla^b\nabla^c\varepsilon = 0, \tag{35}$$

$$\nabla^a\nabla^b\nabla^c\nabla^d\varepsilon = -\frac{1}{2} R^a{}_e{}^b{}_f(R^{-1/2})^{ec}(R^{-1/2})^{fd}, \tag{36}$$

$(R^{-1/2})^{ac}$ is the square root of the inverse matrix of R_{ac} at the origin of the Riemann Normal coordinates.

In the coincidence limit, one finds that, the recursion relation obtained in (27) reduces to

$$n\, a_n - 8R_{\mu\nu}\nabla^{\mu}\nabla^{\nu}a_{n-1} + Qa_{n-1} = 0. \tag{37}$$

To solve this recursion relation, we act on the split equation (27) again with the operator $R_{\mu\nu}\nabla^{\mu}\nabla^{\nu}$, and one gets a equation for $8R_{ab}\nabla^a\nabla^b a_n$ which is solved and substituted in (37). The coefficients a_o and a_1 are solved here.

As per the boundary condition, $a_0 = 1$. In addition, we get the the derivatives of a_0 from the coincidence limit of the equations as explained in [8]

$$\nabla^a a_0 = \frac{1}{2}(R^{-1})^{cd}\nabla^a R_{cd}, \tag{38}$$

$$\begin{aligned} R_{ab}\nabla^a\nabla^b a_0 &= -\frac{1}{4} R_{cd}\left[(R^{-1})^{ab}\nabla^c R_{ab}\right]\left[(R^{-1})^{a'b'}\nabla^d R_{a'b'}\right], \\ &\quad -\frac{1}{2} R_{cd}(R^{-1})^{ab}\nabla^c\nabla^d R_{ab} - R^2. \end{aligned} \tag{39}$$

(In the coincidence limit the indices of the Riemann Normal coordinates can be interchanged with the μ indices.) These equations are used in equation (37) to get a_1. Similarly a a_n for arbitrary n can be obtained. The equation for a_1 is

$$a_1 = 8R_{\mu\nu}\nabla^\mu\nabla^\nu a_0 - Qa_0. \tag{40}$$

Substituting for Q and from (39), we get an expression for a_1 in terms of the curvature invariants and the derivatives of the curvature. This is thus a derivation of the heat kernel expansion of a non-Laplacian operator for non-perturbative gravity. In this derivation of the heat kernel expansion, we have ignored the boundary of the manifold, as we have done a Riemann Normal coordinate expansion about one local point. Inclusion of half-integer t coefficients ensures the inclusion of boundary terms, but that is not relevant for the calculation in this paper.

4.3. The Non-perturbative Effective Action

In the two regimes we considered, $\nabla^\mu R_{\mu\nu} \sim 0$ and $\nabla^\mu R_{\mu\nu} \gg 1$, the terms the zeta function and its derivative are:

$$\zeta(p) = \frac{\mu^{2p}}{16\pi^2\Gamma(p)}\int dt\, t^{p-3}\,\mathrm{Tr}_x e^{-tQ}\sum_i a_i t^i \tag{41}$$

$$\begin{aligned} 16\pi^2\zeta'(p) &= \frac{\mu^{2p}}{\Gamma(p)}\left(\ln\mu^2 - \frac{\Gamma'(p)}{\Gamma(p)}\right)\int dt\, t^{p-3}\,\mathrm{Tr}_x e^{-tQ}\sum_i a_i t^i \\ &+ \frac{\mu^{2p}}{\Gamma(p)}\int dt\, t^{p-3}\ln t\,\mathrm{Tr}_x e^{-tQ}\sum_i a_i t^i. \end{aligned} \tag{42}$$

The finite term as $p \to 0$ appears in the last term of the derivative of the zeta function (as we know this might not be the unique the way to extract the finite term but sure explains a way to cancel the negative term from the classical action). This is given in details in [8]. The finite term (regulator independent) is remarkably proportional to a_1 and is obtained from (42) in [8] (using a regularization $\zeta(0) = -1/2$)

$$\zeta'(0)_{\mathrm{finite}} = -\frac{1}{32\pi^2}\mathrm{Tr}_x a_1 \tag{43}$$

Thus in the effective action $\Gamma_{\text{trace}} + \Gamma_{\text{classical}} = \frac{1}{2}\zeta'(0) + \Gamma_{\text{classical}}$ we get to first order $-\frac{1}{64\pi^2}\,\text{Tr}_x a_1 + \Gamma_{\text{classical}}$. In the subsequent discussion, we fix the a_1 in the weak gravity and strong gravity regimes and find the $\Gamma_{\text{trace}} + \Gamma_{\text{classical}}$.

(i) $\nabla_\mu R_{\mu\nu} \approx 0$, the Faddeev–Popov determinant is

$$\det\big(8(1+2C)6\nabla^4\big) = \det\big(-8(1+2C)6\nabla^2\big)\det(-\nabla^2). \tag{44}$$

In the factorized form, the first scalar determinant is a divergent one for $C < -1/2$, and the second one is a convergent one. We discuss the divergent determinant's contribution to the effective action as this should cancel the divergence from the classical action.

The first scalar determinant is of a Laplacian and thus we use the heat kernel of a Laplacian, and analytically continue to the divergent regime of $C < -1/2$. The a_1 coefficient of the Laplacian is well known, and to quote [11]

$$a_1 = \frac{1}{6}R. \tag{45}$$

What is interesting is that this term is dimensional and it is obvious that the space of measures does not have any length scales like the Planck length to make the term dimensionless in the exponential. To retain the correct dimensions, one had to introduce in the definition of the measure and the determinants a scale. I scale the coefficient a_1 by a appropriate length squared: l_{un}^2 (in [8] this had been taken to be Planck length squared) to get the exponential dimensionless and restore the $16\pi G$ in the classical action. Writing R in terms of $\bar{R}$ and $(\nabla\phi)^2$, and using the constants in the determinant one gets as the coefficient of the kinetic term of the conformal mode in the effective action ($\Gamma_{\text{trace}} + \Gamma_{\text{classical}}$)

$$-\frac{1}{16\pi}\left[1 + \frac{2(1+2C)}{\pi}\frac{l_p^2}{l_{\text{un}}^2}\right]. \tag{46}$$

Thus the positive action takes over at $(1+2C) > -l_{\text{un}}^2\pi/2l_p^2$. From Einstein' action $C = -2$, and Euclidean Einstein gravity has a positive definite effective action for $l_{\text{un}} = l_p$. The number $-\pi/2$ might differ for different regularization schemes, (and conventions for defining determinants from Gaussian integrals) but it is indeed a finite number, and we should be in the realm of a convergent path-integral for Euclidean quantum gravity. Note that the effect of this calculation has been in the end change of the overall sign of the action by a minus sign. This achieves the required 'convergent behavior' of the exponent in the path-integrand. This contribution from the measure has the right sign, irrespective of what the sign of the classical bare action is. We saw that the classical bare action can be unbounded from below, and this change in sign makes those divergent actions contribute with convergent weights to the gravitational path-integral. From a premature analysis of the classical action it seems that the wildly divergent terms appear with a negative sign and thus reversing this sign makes the path-integral potentially convergent.

(ii) In the case of the regime $\nabla^{\mu} R_{\mu\nu} \gg 1$, the relevant operator is (23) and the coefficient a_1 is from (40, 39) (The Q in (40) is absorbed in the exponential of (42)),

$$\begin{aligned} a_1 &= -2R_{\lambda\sigma}[(R^{-1})^{\mu\nu}\nabla^{\lambda}R_{\mu\nu}][(R^{-1})^{\mu'\nu'}\nabla^{\sigma}R_{\mu'\nu'}] \\ &- 4R_{\lambda\sigma}(R^{-1})^{\mu\nu}\nabla^{\sigma}\nabla^{\lambda}R_{\mu\nu} - 8R^2. \end{aligned} \tag{47}$$

We find that $-R^2$ term has exactly the sign required to cancel the contribution from the $(\bar{\nabla}\phi)^2$ term in the classical action, as writing R in terms of $\bar{R}$ and $\bar{\nabla}\phi$, one finds $(\bar{\nabla}\phi)^4$ from R^2, which dominate for configurations with rapidly varying ϕ. The other terms do not give divergent negative terms. Any further negative divergence from $\bar{g}$ is also rendered positive by this squared term. Thus $\Gamma_{\text{trace}} + \Gamma_{\text{classical}}$ in this regime emerges as positive definite. Note in this non-perturbative regime, the effective action at this order does not merely change by an overall minus sign but has additional non-trivial contributions proportional to $\bar{R}^2$ and higher derivative terms, which dominate over the classical bare action.

5. Gauge Fixing and Renormalization

In the previous section we discussed the path-integral with focus on the scalar sector of the metric, without mentioning details about the gauge fixing of the metric. As the diffeomorphism group is factored out of the path-integral, the remaining physical degrees of freedom are specified in a particular gauge. This identifies the equation for the gauge slice in the space of metrics. We derive the gauge fixed path-integral formally in this section. The gauge condition identifies the 'transverse modes' of the graviton. A covariant way of expressing the gauge condition is:

$$F \circ h^{\perp} = 0 \tag{1}$$

e.g. in the case of the orthogonal gauge [7],

$$\nabla^{\mu} h^{\perp}_{\mu\nu} = 0, \quad F \equiv \nabla^{\mu}. \tag{2}$$

Using this, the Fadeev–Popov determinant was obtained in [7] as

$$\det{}_S(16(1+2C))\det{}_V(F \circ F^{\dagger})^{-1/2}\det{}_V(F \circ L) \tag{3}$$

(L is the operator defined in (10). If we confine ourselves to the Landau gauge, $F^{\dagger} = -2\nabla^{\mu}$ the vector determinant of (3) is

$$\det{}_V(-2\nabla \circ L)^{1/2}. \tag{4}$$

Thus the non-perturbative path-integral in orthogonal gauge is

$$\int \mathcal{D}g^{\perp}_{\mu\nu}\mathcal{D}\phi \det{}_V(-2\nabla \circ L)^{1/2} \times \exp\left(\int d^4x\sqrt{\det g^{\perp}}\left[R^{\perp} + (\nabla\phi)^2\right]\right). \tag{5}$$

This integral has to be computed after evaluating the determinant. However as we discussed in the previous sections, the scalar sector of the measure can be isolated even prior to gauge fixing. We see whether this agrees with the computation of the scalar determinant in the gauge fixed determinant (4). This vector determinant is equivalent to the inverse of

$$\int \mathcal{D}\xi e^{-\langle \xi F \circ L \xi \rangle}, \tag{6}$$

where ξ is a vector field. In case we write the $\xi^{\mu} = \hat{\xi}^{\mu} + \nabla^{\mu}\tilde{\phi}$, we can isolate a scalar determinant in the above

$$\int \mathcal{D}\tilde{\phi}\det{}_S(-\nabla^2)^{1/2} e^{-2\int d^4x \tilde{\phi}\nabla^{\mu}\nabla^{\nu}L_{\mu\nu\lambda}\nabla^{\lambda}\tilde{\phi}}. \tag{7}$$

The scalar sector of the Fadeev–Popov determinant is thus

$$\det_S(-\nabla^2)^{-1/2} \times \det_S\Big[8(1+2C)\left(-2(\nabla)^4 + 4\nabla_\mu\nabla^2\nabla^\mu + 4\nabla_\mu\nabla_\rho\nabla^\mu\nabla^\rho\right)\Big]^{1/2}. \quad (8)$$

As is obvious the scalar sector of the determinant is precisely the determinant obtained in (3), which thus confirms our 'isolation' of the scalar determinant as correct.

This thus shows that the effective action is indeed positive definite, and the resolution obtained in this paper can be used to render any negative action into a positive action. Question however remains whether the resultant positive definite action can give computable amplitudes, and if so can the path-integral be renormalized at a non-perturbative level. However, we cannot use the techniques of perturbative field theoretic renormalization or for that matter 'scattering amplitudes' to calculate in this approach. The clues come from recent computations of the quantum amplitude using CDT. And we discuss some of the results here.

5.1. A Non-perturbative Finite Path-integral

In QCD one uses Lattice methods in the non-perturbative regime of this theory. The field is described on discretised links and vertices which comprise a Lattice. For gravity too, the Lattice methods give a way of regulating otherwise divergent quantities. The Causal Dynamical Triangulation uses a piecewise flat discretization and the curvature of the space-time lives at the vertices. These piece-wise flat building blocks are tetrahedrons and using them a entire 'discretised' four geometry can be built. The Einstein action is written in terms of the building blocks of the discrete space-time, and is thus a function of the number of building blocks and the edge lengths of each tetrahedron [14]. The CDT approach taken by Ambjorn, Jurkiewicz, Loll (AJL) and collaborators [5] use a 'Lorentzian' discrete space-time and build a Euclidean sum over geometries obtained by Wick Rotation. This restricts the Euclidean geometries to those which do not have singular lightcones, and surprisingly in the continuum limit, the path-integral is finite. It is believed that Causality acts as a regulator. We briefly describe this CDT computation in this section.

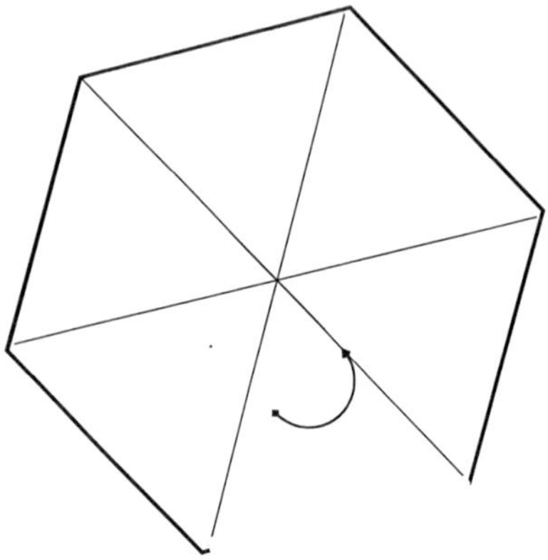

Figure 3. An example of 2-dimensional triangulation.

To understand discretization of curved space-time, a two dimensional example is a useful illustration, in Figure 3 it is shown that if one takes equilateral triangles to fill flat space, then six triangles meet at a vertex. If we remove one triangle, and identify the edges of the remaining triangles bounding the empty space, then the resultant space-time is a 'cone' with a curvature at the vertex proportional to the deficit angle. If the number of 'removed' triangles is increased, the deficit angle increases and also the curvature at the cone vertex. The two dimensional curvature is measured to be proportional to the deficit angle. Thus curved space-time has the 'vertices' as the points at which the curvature lives. Incase the edge lengths are different for the three sides of a triangle, the number of triangles meeting at a vertex will differ, and the deficit angle will be different. e.g. In case the two dimensional space-time has Lorentzian signature, as considered in the AJL model, the sum over all angles around a vertex with j incoming and k outgoing time like edges is $2\pi + (4 - j - k)\cos^{-1}(3/2)$. Flat Minkowski space corresponds to j=k=2 at all vertices, as obviously the total angle is 2π and the deficit angle or the curvature is zero.

Once the simplices are defined, the set of simplices linked at the vertices comprise a discrete geometry. The path-integral is a sum over these discrete geometries given the boundaries. The sum might be defined by varying the edge lengths, or by keeping the edges fixed and varying the number of simplices or the 'triangulation'. The AJL model of the gravitational path-integral uses a proper-time foliation of space-time. This is the case where one direction is chosen as time and discrete piece-wise flat building blocks (simplices) are used to fill in between the time slices. This use of proper-time naturally includes causality restrictions (interpreted as a 'gauge-fixed' version of the gravity path-integral). In four dimensions, the regularized Lagrangian is written in terms of number of building blocks which link one time slice to another. The simplices are of the following types (i) One Vertex on one slice and four on the next (1,4) or the reverse of this (4,1) and (ii) one 'triangular face' or three vertices on one slice and one link or two vertices on the other slice (3,2) and the time reversed of this (2,3). In both the above simplices some of the faces which lie on the spatial slice are space-like, and those linking two space-like slices are time like, or they are bounded by edges which have 'negative length'.

In CDT the Einstein action uses the cosmological constant.

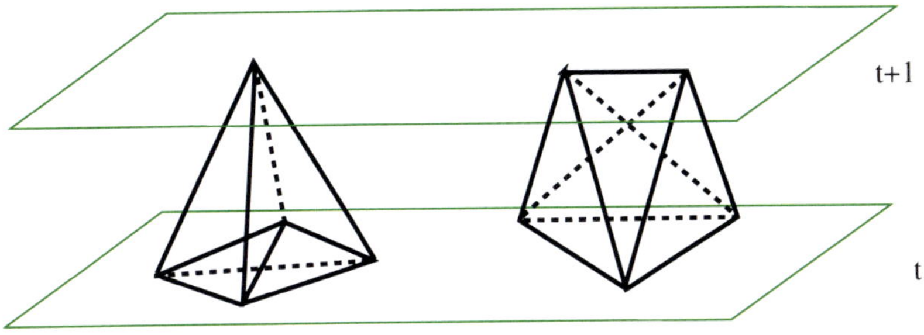

Figure 4. The simplices of CDT: Copyright J. Ambjorn, J. Jurkiewicz, R. Loll.

$\int d^4x\sqrt{-g}(R-2\Lambda)$ gets converted to upon discretization

$$S^{\text{Regge}} = k\left[\sum_{\text{space}} V_s \frac{1}{\iota}\left(2\pi - \sum\Theta\right) + \sum_{\text{time}} V_t(2\pi - \sum\Theta)\right] - \lambda\left(\sum V(4,1) + \sum V(3,2)\right), \quad (9)$$

$V_{s/t}$ stands for time/ space-like Volumes of the faces which meet at a vertex and $V(4,1)$ and $V(3,2)$ are the four volumes of the simplices. $k = 1/16\pi G$,

$l = \Lambda/G$. Θ is the sum total of the diehedral angles meeting at a vertex. The volume of the simplices can be calculated and after some algebra and a suitable Lorentzian to Euclidean conversion this becomes

$$S_E(T) = -(\kappa_0 + 6\Delta)N_0 + \kappa_4(N_4^{(4,1)} + N_4^{(3,2)}) + \Delta(2N_4^{(4,1)} + N_4^{(3,2)}). \quad (10)$$

The k_0, Δ, k_4 are functions of the parameters of the theory and define the phase diagram of the path-integral. N_0 is the total number of vertices, $N_4^{(4,1)/(3,2)}$ is the total number of the simplices of the (4,1) and (3,2) types. T stands for a particular triangulation or discretization. Thus given a triangulation, the action is fixed and the partition function is a sum over the number of triangulations of the four geometry.

The partition function is

$$Z(\kappa_0, \kappa_4, \Delta) = \sum_T \frac{1}{C_T} e^{-S_E(T)}, \quad (11)$$

where T is a discretization and C_T is a combinatorial factor. The CDT is a sum over discretizations or the number of ways to fill the space-time with simplices whose volume per simplex is fixed. Numerical simulations and probing the sum using various approximations, it emerges that there are three phases of the system. These phases are characterized using the volume of three time slices (i) The phase A has three geometries with volumes distributed along the time direction like 'pearls on a string'. These 'small universes' are linked by three time slices with the minimal volume allowed by the discretization i.e. five tetrahedrons glued together. (ii)The phase B does not have the time direction, and there is only one spatial slice which has a sizable volume. (iii) Phase C seems to have emergent space-times which seemingly appear 'real'. The time extent scales as $V^{1/4}$ and the spatial volume as $V^{3/4}$ which is nearest to classical geometries observed in the Universe.

This emergent universe is one of the first derivations of a quantum gravity universe. There are several other results which are derived using the Montecarlo simulated universe, including a average expectation value of the three volume $< V_3(t) >$, which agrees with the solution one would obtain from a classical action of the minisuperpace cosmological scale factor, however with a sign reversed Kinetic term. This also confirms the expectations of a quantum effective

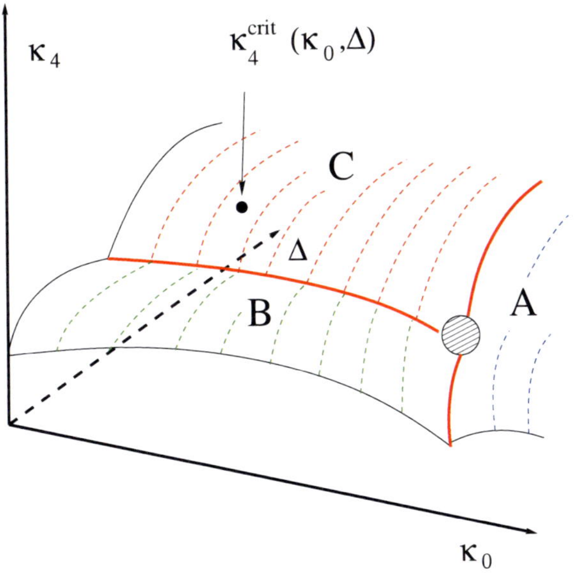

Figure 5. The phases of CDT quantum gravity copyright J. Ambjorn, J. Jurkiewicz, R. Loll.

action which is positive definite [8]. In terms of regularized and normalized constants the volume function is

$$\sqrt{g_{tt}} V_3^{\mathrm{cl}}(t) = V_4 \frac{3}{B} \cos^3\left(\frac{t}{B}\right), \tag{12}$$

where t is the time and B is constant related to the total volume, V_{cl} is the classical three volume and $V_4(t) = \int dt \sqrt{g_{tt}} V_3(t)$ is the integral over the three volumes of the slices. (12) is precisely the volume profile expected from a classical universe derived form a reversed signature action. Another remarkable result is the observation that at short distances the universe has Hausdorff dimension 2 and appears fractal [5]. This agrees with the results of the Reuter et al in this volume in Chapter 2.

6. Conclusion

In this article we introduced the path-integral for gravity, provided evidence that the effective action is positive definite. We discussed a non-perturbative CDT computation which gives useful predictions about quantum gravity some of which are confirmed from other approaches like the fractal behavior of space-time. In Section 2 we introduced the path-integral for gravity as a generalization of the path-integral for a particle. We showed that unlike the case of particle, the measure in the gravitational path integral is non-trivial due to diffeomorphism invariance. This measure is obtained through the Faddeev–Popov procedure and is obtained as a functional determinant. After defining the path-integral we discuss the infinities which might occur in the computation, including the unboundedness from below of the Euclidean action and the non-renormalizability of perturbative amplitudes. In the next Section 4 we show how the measure can be used to cure the potential problem due to the unboundedness of the Euclidean action. This result is very promising for the computability of the path-integral. In Section 5 we donot perform the sum over histories as the continuum techniques cannot be used. Instead we discuss results from CDT which uses Lattice techniques to obtain interesting results, including an emergent semiclassical universe and fractal behavior of the emergent space-time at short distances. Whereas there can be debate about the approximations and the validity of the results, it is obvious that quantum gravity is computable using existing techniques. However, experimental verification will confirm whether these theoretical computations are a correct way of describing nature. Therefore in the next few years quantum gravity phenomenology should acquire importance, including efforts to find signatures of quantum gravity in astrophysical experiments. I have used chaos in black hole orbits to magnify quantum gravity effects [15], an exact experiment is yet to be designed. This is work in progress. When experimental results emerge, we will have a thorough understanding of the four interactions of nature. Whether there will be a unified theory to describe the interactions in a unique way however remains a mystery. Why does nature obey these particular symmetries and the principle of least action would continue to puzzle future generations.

References

[1] R. P. Feynman, *Rev. Mod. Phys.* 20 367 (1930).

[2] B. S. DeWitt, *Phys. Rev.* 162 1239 (1967).

[3] G. 'tHooft, M. Veltman, *Ann. Inst. Henri. Poincare* 20 69 (1974).

[4] G. W. Gibbons, S. W. Hawking, M. J. Perry, *Nucl. Phys. B* 138 141 (1978).

[5] J. Ambjorn, A. Gorlich, J. Jurkiewicz, R. Loll '*CDT-an Entropic Theory of Quantum Gravity*' arXiv:1007.2560 [hep-th]

[6] C. Rovelli, '*Zakopane Lectures on Loop Quantum Gravity*' arXiv:1102.3660 [gr-qc] and references therein.

[7] Z. Bern, M. Blau, E. Mottola, *Phys. Rev. D* 43 1212 (1991).

[8] A. Dasgupta, *General. Relativ. Gravit.* 43 2237 (2011). [gr-qc:0801.04770]

[9] C. Itzykson, J-B. Zuber, '*Quantum Field Theory*' Dover Publications (2006), S. Weinberg, '*Quantum Theory of Fields*' Cambridge University Press (2005).

[10] P. Mazur, E. Mottola, *Phys. Rev. D* 64 104022 (2001), *Nucl. Phys. B* 341 187 (1990).

[11] B. S. Dewitt, *Dynamical theory of groups and fields*, Gordon and Breach Science Publishers, (1965).

[12] A. O. Barvinsky, G. A. Vilkovisky, *Phys. Rept.* 119 1 (1985).

[13] A. Dasgupta, R. Loll, *Nucl. Phys. B* 606 357 (2001). A. Dasgupta, *JHEP* 0207 062 (2002).

[14] T. Regge, *Nuovo. Cimento.* 19 558 (1961).

[15] A. Dasgupta, *JCAP* 1005:011 (2010).

INDEX

A

amplitude, 22, 111, 116, 122, 123, 124, 128, 130, 131, 138, 139, 140, 147, 161
anisotropy, 131
annihilation, 23
argon, 49
atoms, vii, 2, 10

B

baggage, 9
Big Bang, 25
births, 131
black hole, 3, 4, 5, 8, 9, 23, 25, 42, 89, 95, 96, 97, 98, 99, 102, 103, 108, 110, 111, 115, 131, 166
black hole entropy, 96, 97
boils, 60, 62
boson, 97
bounds, 23, 26
breakdown, 25
BTZ black hole, 99
building blocks, 2, 161, 163

C

cation, 61
causality, 9, 29, 31, 34, 163
CERN, 38
chaos, 166
classes, 9, 28, 75
community, 4, 35
complexity, 76
complications, 2, 11, 31, 53, 127
computation, 37, 70, 89, 131, 138, 142, 145, 147, 160, 161, 166
computing, 45, 132, 145, 147
conception, 132
conference, 38, 134, 135
configuration, 19, 63
conjugation, 129
connectivity, 8
conservation, 3, 148
construction, 7, 10, 11, 13, 19, 21, 22, 23, 31, 32, 33, 34, 42, 43, 44, 52, 56, 58, 73, 74, 75, 87, 106
contour, 118, 119, 120, 122, 123, 126, 132, 145
copyright, 3, 163, 165
cosmic rays, 35
cosmos, 116, 127, 128, 129, 130, 131, 133
coupling constants, 29, 84
covering, 16, 17, 116
criticism, 23, 33

cure, 34, 166
cures, 146

D

damping, 125, 130, 131, 132, 133
dark energy, 1
dark matter, 1, 35
decay, 124
decomposition, 23, 45, 53, 69, 70, 84
decoupling, 70
deficit, 162
deformation, 118, 119, 126
degenerate, 134
derivatives, 47, 57, 60, 63, 65, 68, 70, 95, 97, 100, 104, 146, 157
deviation, 83, 86
differential equations, 49, 63, 66
diffusion, 26, 153
diffusion process, 26
dimensionality, 72, 73, 78
discreteness, 10, 12, 14, 23, 24, 25, 33, 35
discretization, 105, 108, 109, 161, 162, 163, 164
dispersion, 89
distribution, 11, 12, 14, 24, 26, 104
divergence, 98, 129, 144, 145, 146, 148, 149, 152, 158, 159

E

effective field theory, 42
electron, 4
emission, 122
energy, 4, 9, 24, 26, 81, 101, 105, 108, 122, 123, 124, 126, 127, 129
energy density, 101
entropy, 23, 28, 32, 33, 34, 96, 97, 110, 111
equality, 47
equilibrium, 132
Euclidean space, 43
evaporation, 42
evidence, 1, 13, 41, 42, 87, 116, 166
evolution, 2, 24, 26, 27, 31, 51, 59, 60, 61, 68, 69, 140
exercise, 32, 34
extraction, 42

F

fears, 33
fermions, 35, 88
Feynman diagrams, 138, 142, 147, 148, 149
field theory, 49, 87, 97, 105, 137, 141
financial, 89
financial support, 89
fine tuning, 88
fluctuations, 2, 20, 21, 24, 42, 70, 152
fluid, 10
formation, 2, 16, 27, 58, 95, 103, 143, 144
formula, 117, 125
fractal dimension, 42, 89
fractal properties, 89
France, 41, 135
freedom, 23, 69, 96, 97, 99, 104, 143, 145, 159

G

gauge invariant, 54
gauge theory, 10, 53, 54
General Relativity, 5, 42, 88, 91, 134
geometry, 8, 9, 10, 18, 27, 33, 42, 64, 88, 89, 116, 128, 161, 162, 164
Germany, 41
gluons, 149
graph, 18, 109
gravitational field, 57
gravitational force, 2
gravity, vii, 2, 4, 5, 7, 9, 24, 28, 33, 41, 42, 43, 53, 54, 59, 63, 69, 70, 71, 74, 87, 88, 91, 95, 96, 97, 98, 99, 101, 102, 103, 104, 110, 116, 120, 128, 130, 138, 142, 147, 150, 151, 153, 154, 157, 158, 161, 163, 166
growth, 7, 28, 29, 30, 31, 34, 131
growth dynamics, 7, 34

growth models, 28, 29, 31
guilty, 130

H

Hamiltonian, 4, 24, 28, 95, 96, 97, 100, 101, 103, 104, 105, 106, 107, 108, 111, 130
Hausdorff dimension, 2, 165
Hawking radiation, 34, 88
height, 15, 21
Hilbert space, 31, 70
history, 1, 138, 139
hospitality, 36, 112
host, 8
human, 2
hybrid, 125, 133

I

ideal, 42, 88, 96, 110
identification, 13, 144
independence, 51, 76, 87
India, 7
inevitability, 96
infancy, 34
ingredients, 33, 58
inhomogeneity, 131
integration, 45, 47, 54, 58, 60, 67, 69, 73, 104, 119, 120, 122, 123, 124, 126, 132
interference, 31
invariants, 62, 63, 64, 68, 76, 82, 83, 87, 157
isolation, 150, 161
Israel, 91
issues, 32, 89, 96, 123, 130, 138, 142

J

Jordan, 22, 23
justification, 76

L

landscape, 28, 32
laws, 1, 4
lead, 2, 21, 55, 56, 127, 130
legs, 149
light, 3, 14, 33, 120
linear function, 148

M

magnitude, 24, 82, 119
man, 1
manifolds, 28, 146
Maryland, 134
mass, 3, 26, 43, 44, 45, 48, 49, 53, 56, 78, 95, 97, 99, 101, 103, 104, 108, 109, 110, 125, 127, 129, 132, 140
matrix, 18, 19, 22, 23, 34, 46, 53, 60, 72, 80, 81, 83, 156
matter, vii, 2, 3, 11, 26, 35, 45, 65, 70, 88, 98, 100, 101, 103, 104, 111, 161
measurement, 31, 34
meth, 2
Minkowski spacetime, 11, 12, 14, 16, 17, 19, 22, 26, 27, 28
models, 4, 23, 26, 28, 31, 34, 35, 96, 98, 110
modifications, 46
modulus, 139
molecular structure, 10
momentum, 25, 26, 42, 44, 45, 46, 52, 57, 64, 75, 103, 104, 106, 147, 148
Monte Carlo method, 32
motivation, 47
MRI, 133

N

neglect, 127, 128
nerve, 17
normalization constant, 139
null, 14, 96, 97, 101, 102

O

obstruction, 120
ordinary differential equations, 51, 62
organ, 125
organize, 54

P

Pacific, 37
parallel, 119, 120
particle mass, 45
particle physics, 1, 42
partition, 32, 129, 151, 152, 164
path integrals, 116
pathology, 145
percolation, 29, 30, 31
phenomenology, 12, 89, 166
Philadelphia, 36, 134
physical features, 50
physical laws, 27
physics, vii, 1, 2, 3, 4, 5, 10, 21, 28, 33, 42, 51, 54, 56, 88, 96, 115, 133
polymer, 28, 95, 97, 111
prejudice, 88
preparation, 36, 38, 39, 112
principles, 29, 33, 34
probability, 11, 12, 25, 26, 30, 31, 33, 130, 139, 142
probability distribution, 26
project, 50
propagation, 2, 27
protons, 26

Q

QCD, 78, 161
QED, 5, 87
quantization, 42, 53, 70, 88, 95, 97, 105, 111, 137, 138
quantum cosmology, 30, 31, 116, 130, 133, 134
quantum dynamics, 13, 23, 31, 32, 34
quantum field theory, 22, 23, 24, 61, 73, 87, 88, 138, 139, 147
quantum gravity, vii, 1, 2, 4, 5, 9, 23, 24, 25, 32, 33, 34, 35, 41, 42, 43, 69, 87, 88, 96, 97, 99, 116, 133, 137, 158, 164, 165, 166
quantum Hall effect, 105
quantum mechanics, 111, 116, 133
quantum structure, 97
quantum theory, vii, 2, 5, 31, 71, 88, 111
query, 13, 24

R

radius, 21, 63, 64, 101, 102, 128, 129, 130, 133
rash, 116
reading, 131
real numbers, 140
reasoning, 69
recall, 116
reflexivity, 8
relatives, 133
relativity, 2
relevance, 16, 42
reliability, 67, 74, 78
renormalization, 41, 43, 48, 52, 62, 63, 71, 161
requirements, 48, 51
researchers, vii, 33
resolution, 89, 95, 96, 150, 152, 161
response, 23
restrictions, 33, 163
Riemann tensor, 63
risk, 133
root, 103, 106, 111, 118, 119, 125, 130, 147, 152, 156
Royal Society, 36
rules, 10, 29, 32

S

safety, vii, 4, 41, 42, 43, 71, 73, 74, 75, 77, 84, 86, 87, 88, 89, 91

scalar field, 7, 19, 22, 43, 50, 98
scalar field theory, 43
scaling, 15, 78, 81, 83, 155
scattering, 59, 105
scent, 118, 119
scope, 31, 152
senses, 96
set theory, vii, 7, 9, 12, 13, 24, 28, 33, 34, 35
shape, 44, 67, 68, 70, 74, 77, 79, 80, 81, 82, 83, 84, 86, 128
showing, vii, 22
signals, 27
signs, 83, 121
simulation, 5
simulations, 16, 18, 24, 32, 34, 89, 126, 164
Singapore, 91
solution, 25, 54, 70, 73, 77, 89, 99, 119, 124, 146, 154, 155, 164
spacetime, 7, 8, 9, 10, 11, 12, 13, 14, 15, 16, 17, 18, 19, 21, 22, 23, 24, 25, 26, 28, 29, 30, 32, 33, 35, 48, 78, 88, 89, 97, 98, 99, 100, 101, 102, 111, 115, 116, 128, 129, 133
spectral dimension, 89
spin, 35, 63, 70, 142
stability, 77, 80, 86
state, 23, 33, 108
states, 12, 26, 97, 105, 106, 107, 108, 109, 111, 132, 146
stochastic model, 24, 25
string theory, 5, 96
strong force, 1
structure, 2, 4, 8, 9, 12, 28, 32, 35, 36, 43, 57, 62, 69, 70, 88, 95, 131
structure formation, 35
style, 46
substitutions, 61, 99, 129
supersymmetry, 88
suppression, 44, 57
symbolism, 126
symmetry, 23, 33, 42, 48, 51, 52, 97, 105, 110, 111, 129, 131, 142, 152

T

techniques, 42, 63, 66, 97, 138, 145, 152, 166
temperature, 42
tensor field, 53, 58, 65
testing, 51, 74, 76
textbooks, 116
thermodynamics, 5, 115
topological invariants, 17, 18
topological structures, 14
topology, 8, 9, 13, 16, 18, 32, 106, 109, 130, 134
total energy, 101
trajectory, 49, 51, 71, 73, 75, 81, 82, 86, 122, 123, 125
transformation, 47, 55, 58, 104, 143, 144, 149
transformations, 9, 54, 56, 58, 59, 106, 143, 149
translation, 106
transport, 120
treatment, 48, 69, 74, 123, 132, 146
triangulation, 16, 89, 162, 163, 164
tunneling, 115, 116, 122, 128, 129, 131, 132, 133

U

unification, 88
uniform, 26, 33
unions, 30, 31
universality, 63, 74, 76, 78, 83
universe, 2, 5, 24, 25, 28, 30, 128, 130, 131, 164, 165, 166

V

vacuum, 23, 95, 99, 104, 110, 111
variables, 58, 59, 104, 107, 110, 126, 145
variations, 82, 83, 118, 124
vector, 26, 31, 35, 48, 49, 50, 53, 54, 65, 70, 71, 87, 99, 143, 144, 149, 156, 160
velocity, 71

W

water, 110
workers, 34
worry, 24, 129

Y

yield, 16, 23, 25, 45, 77, 109